ζ Zeta

Decimals and Percents

Instruction Manual

by Steven P. Demme

1-888-854-MATH (6284) - mathusee.com
sales@mathusee.com

Zeta Instruction Manual: Decimals and Percents
©2012 Math-U-See, Inc.
Published and distributed by Demme Learning

All rights reserved. No part of this book may be reproduced, stored in a retrieval system, or transmitted in any form by any means—electronic, mechanical, photocopying, recording, or otherwise—without prior written permission from Demme Learning.

mathusee.com

1-888-854-6284 or +1 717-283-1448 | demmelearning.com
Lancaster, Pennsylvania USA

ISBN 978-1-60826-206-9
Revision Code 0222-C

Printed in the United States of America by CJK Group
    3 4 5 6 7 8 9 10

For information regarding CPSIA on this printed material call: 1-888-854-6284 and provide reference #0222-04042024

Building Understanding in Teachers and Students to Nurture a Lifelong Love of Learning

At Math-U-See, our goal is to build understanding for all students.

We believe that education should be relevant, skills-based, and built on previous learning. Because students have a variety of learning styles, we believe education should be multi-sensory. While some memorization is necessary to learn math facts and formulas, students also must be able to apply this knowledge in real-life situations.

Math-U-See is proud to partner with teachers and parents as we use these principles of education to **build lifelong learners.**

Curriculum Sequence

∫ **Calculus**

cos **PreCalculus** with Trigonometry

xy **Algebra 2**

Δ **Geometry**

x² **Algebra 1**

x **Pre-Algebra**

ζ **Zeta** Decimals and Percents

ε **Epsilon** Fractions

δ **Delta** Division

γ **Gamma** Multiplication

β **Beta** Multiple-Digit Addition and Subtraction

α **Alpha** Single-Digit Addition and Subtraction

ρ **Primer** Introducing Math

Math-U-See is a complete, K-12 math curriculum that uses manipulatives to illustrate and teach math concepts. We strive toward "Building Understanding" by using a mastery-based approach suitable for all levels and learning preferences. While each book concentrates on a specific theme, other math topics are introduced where appropriate. Subsequent books continuously review and integrate topics and concepts presented in previous levels.

Where to Start
Because Math-U-See is mastery-based, students may start at any level. We use the Greek alphabet to show the sequence of concepts taught rather than the grade level. Go to mathusee.com for more placement help.

Each level builds on previously learned skills to prepare a solid foundation so the student is then ready to apply these concepts to algebra and other upper-level courses.

Major concepts and skills for Zeta:
- Expanding understanding of place value from positive powers of ten to include decimals
- Fluently adding, subtracting, multiplying, and dividing multiple-digit decimals using place-value strategies
- Solving real-world problems with decimals and percents

Additional concepts and skills for Zeta:
- Understanding and simplifying exponents
- Understanding negative numbers and representing them on the coordinate plane
- Using properties of operations to simplify and evaluate algebraic expressions
- Interpreting and graphing relationships between dependent and independent variables
- Understanding plane geometry and geometric symbols
- Using ratio reasoning to solve problems
- Measuring statistical variability and distributions

Find more information and products at mathusee.com

Contents

	Curriculum Sequence	4
	How To Use	7
Lesson 01	Exponents; Word Problem Tips	11
Lesson 02	Place Value	15
Lesson 03	Decimal Numbers with Expanded Notation	19
Lesson 04	Add Decimal Numbers	23
Lesson 05	Subtract Decimal Numbers	27
Lesson 06	Metric System Origin–Meter, Liter, Gram	29
Lesson 07	Metric System–Latin Prefixes	33
Lesson 08	Metric System Conversion–Part 1	37
Lesson 09	Multiply by 1/10 or 0.1	41
Lesson 10	Multiply Decimals by 1/100 or 0.01	45
Lesson 11	Finding a Percentage	49
Lesson 12	Finding a Percentage > 100%	53
Lesson 13	Reading Percentages in a Pie Graph	57
Lesson 14	Multiply All Decimals	59
Lesson 15	Metric System Conversions – Part 2	61
Lesson 16	Computing Area and Circumference	65
Lesson 17	Dividing a Decimal by a Whole Number	69
Lesson 18	Dividing a Whole Number by a Decimal	73
Lesson 19	Solving for an Unknown	77
Lesson 20	Dividing a Decimal by a Decimal	79
Lesson 21	Decimal Remainders	81
Lesson 22	More Solving for an Unknown	85
Lesson 23	Convert Any Fraction	87
Lesson 24	Decimals as Rational Numbers	91
Lesson 25	Mean, Median, and Mode	95
Lesson 26	Probability	97
Lesson 27	Points, Lines, Rays, and Line Segments	99
Lesson 28	Planes and Symbols	101
Lesson 29	Angles	103
Lesson 30	Types of Angles	105
	Student Solutions	107
	Application and Enrichment Solutions	187
	Test Solutions	195
	Symbols and Tables	219
	Glossary	221
	Master Index for General Math	227
	Zeta Index	229

HOW TO USE

Five Minutes for Success

Welcome to *Zeta*. I believe you will have a positive experience with the unique Math-U-See approach to teaching math. These first few pages explain the essence of this methodology, which has worked for thousands of students and teachers. I hope you will take a few minutes and read through these steps carefully.

I am assuming your student has a thorough grasp of the four basic operations (addition, subtraction, multiplication, and division) and a mastery of fractions.

If you are using the program properly and still need additional help, you may visit us online at mathusee.com or call us at 888-854-6284. –**Steve Demme**

The Goal of Math-U-See

The underlying assumption or premise of Math-U-See is that the reason we study math is to apply math in everyday situations. Our goal is to help produce confident problem solvers who enjoy the study of math. These are students who learn their math facts, rules, and formulas and are able to use this knowledge to solve word problems and real-life applications. Therefore, the study of math is much more than simply committing to memory a list of facts. Our program includes memorization, but it also encompasses learning the underlying concepts of math that are critical to successful problem solving.

Support and Resources

Math-U-See has a number of resources to help you in the educational process.

Many of our customer service representatives have been with us for over 10 years. They are able to answer your questions, help you place your student in the appropriate level, and provide knowledgeable support throughout the school year.

Visit mathusee.com to use our many online resources, find out when we will be in your neighborhood, and connect with us on social media.

More than Memorization

Many people confuse memorization with understanding. Once while I was teaching seven junior high students, I asked how many pieces they would each receive if there were fourteen pieces. The students' response was, "What do we do: add, subtract, multiply, or divide?" Knowing how to divide is important, but understanding when to divide is equally important.

The Suggested 4-Step Math-U-See Approach

In order to train students to be confident problem solvers, here are the four steps that I suggest you use to get the most from the Math-U-See curriculum.

Step 1. Prepare for the lesson
Step 2. Present and explore the new concept together
Step 3. Practice for mastery
Step 4. Progress after mastery

Step 1. Prepare for the lesson

Watch the video lesson to learn the new concept and see how to demonstrate this concept with the manipulatives when applicable. Study the written explanations and examples in the instruction manual.

Step 2. Present and explore the new concept together

Present the new concept to your student. Have the student watch the video lesson with you, if you think it would be helpful. The following should happen interactively.

a. **Build:** Use the manipulatives to demonstrate and model problems from the instruction manual. If you need more examples, use the appropriate lesson practice pages.

b. **Write:** Write down the step-by-step solutions as you work through the problems together, using manipulatives.

c. **Say:** Talk through the why of the math concept as you build and write.

Give as many opportunities for the student to "Build, Write, Say" as necessary until the student fully understands the new concept and can demonstrate it to you confidently. One of the joys of teaching is hearing a student say *"Now I get it!"* or *"Now I see it!"*

Step 3. Practice for mastery

Using the lesson practice problems from the student workbook, have students practice the new concept until they understand it. It is one thing for students to watch someone else do a problem; it is quite another to do the same problem themselves. Together, complete as many of the lesson practice pages as necessary (not all pages may be needed) until the student understands the new concept, demonstrating confident mastery of the skill. Remember, to demonstrate mastery, your student should be able to teach the concept back to you using the Build, Write, Say method. Give special attention to the word problems, which are designed to apply the concept being taught in the lesson. If your student needs more assistance, go to mathusee.com to find review tools and other resources.

Step 4. Progress after mastery

Once mastery of the new concept is demonstrated, advance to the systematic review pages for that lesson. These worksheets review the new material as well as provide practice of the math concepts previously studied. If the student struggles, reteach these concepts to maintain mastery. If students quickly demonstrate mastery, they may not need to complete all of the systematic review pages.

In the 2012 student workbook, the last systematic review page for each lesson is followed by a page called "Application and Enrichment." These pages provide a way for students to review and use their math skills in a variety of different formats. Some of the Application and Enrichment pages introduce terms and ideas that a student may encounter on standardized tests. Mastery of these concepts is *not* necessary in order to move to the next level of Math-U-See. You may decide how useful these activity pages are for your particular student.

Now you are ready for the lesson tests. They were designed to be an assessment tool to help determine mastery, but they may also be used as extra worksheets. Your student will be ready for the next lesson only after demonstrating mastery of the new concept and maintaining mastery of concepts found in the systematic review worksheets.

Tell me, I forget. Show me, I understand. Let me do it, I remember.
 –Ancient Proverb

To this Math-U-See adds, *"Let me teach it, and I will have achieved mastery!"*

Length of a Lesson

How long should a lesson take? This will vary from student to student and from topic to topic. You may spend a day on a new topic, or you may spend several days. There are so many factors that influence this process that it is impossible to predict the length of time from one lesson to another. I have spent three days on a lesson, and I have also invested three weeks in a lesson. This experience occurred in the same book with the same student. If you move from lesson to lesson too quickly without the student demonstrating mastery, the student will become overwhelmed and discouraged as he or she exposed to more new material without having learned previous topics. If you move too slowly, the student may become bored and lose interest in math. I believe that as you regularly spend time working along with the student, you will sense the right time to take the lesson test and progress through the book.

By following the four steps outlined above, you will have a much greater opportunity to succeed. Math must be taught sequentially, as it builds line upon line and precept upon precept on previously-learned material. I hope you will try this methodology and move at the student's pace. As you do, I think you will be helping to create a confident problem solver who enjoys the study of math.

LESSON 1

Exponents; Word Problem Tips

Multiplication is a shortcut for fast adding of the same number. To take this a step further, a shortcut for fast multiplying of the same number is raising to a power, which is represented with *exponents.* You can think of exponents in several ways. Picture a square that is 10 over and 10 up. It is 10 two ways (over and up); it is also a square, or "ten squared." By definition, the exponent—in this case 2—stands for how many times 10 is used as a factor. The number 10 is used as a factor twice.

Another way to state it is "ten to the second power" or "ten to the power of two." Any number may be expressed as a square, such as "two squared" or "three squared." Figure 1 shows several ways to express the same idea.

Figure 1

5
5

3
3

5 over and 5 up
5 used as a factor 2 times
5 to the second power
5 squared
5 to the power 2
5 raised to the power of 2

$5^2 = 5 \cdot 5 = 25$

3 over and 3 up
3 used as a factor 2 times
3 to the second power
3 squared
3 to the power 2
3 raised to the power of 2

$3^2 = 3 \cdot 3 = 9$

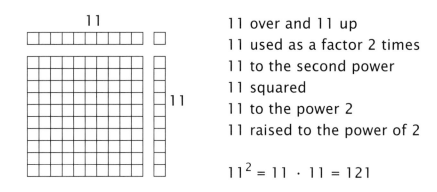

11 over and 11 up
11 used as a factor 2 times
11 to the second power
11 squared
11 to the power 2
11 raised to the power of 2

$11^2 = 11 \cdot 11 = 121$

While it is difficult to show them with the manipulatives, you can have numbers raised to other powers than two. An example is $2 \times 2 \times 2 = 2^3$, which equals 8. Another example is $3 \cdot 3 \cdot 3 \cdot 3 = 3^4$. It helps to say, "Three used as a factor four times is the same as 3^4, which is 81."

Example 1
$7^2 = 7 \cdot 7 = 49$

Example 2
$10 \times 10 \times 10 \times 10 \times 10 = 10^5 = 100{,}000$

When a number is raised to a power of three (for example, 4^3), it may be read as "four to the third power" or "four cubed." You can make a figure that is a cube with dimensions 4 by 4 by 4.

Figure 2

4 over and 4 up and 4 deep
4 used as a factor 3 times
4 to the third power
4 cubed
4 to the power 3
4 raised to the power of 3

$4^3 = 4 \cdot 4 \cdot 4 = 64$

Word Problem Tips

Parents often find it challenging to teach children how to solve word problems. Here are some suggestions for helping your student learn this important skill.

The first step is to realize that word problems require both reading and math comprehension. Don't expect a child to be able to solve a word problem if he does not thoroughly understand the math concepts involved. On the other hand, a student may have a math skill level that is stronger than his or her reading comprehension skills. Below are a number of strategies to improve comprehension skills in the context of story problems. You should decide which ones work best for you and your child.

Strategies for word problems

1. Ignore numbers at first and read the story. It may help some students to read the question aloud. Every word problem tells a story. Before deciding what math operation is required, let the student retell the story in his own words. Who is involved? What are they doing? For example, are they receiving gifts, losing something, or dividing a treat?

2. Relate the story to real life, perhaps by using names of family members. For some students, this may make the problem more interesting and relevant.

3. Build, draw, or act out the story. Use the blocks or actual objects when practical. Especially in the lower levels, you may require the student to use the blocks for word problems even after the facts have been learned. Don't be afraid to use a little drama as well. The purpose is to make it as real and meaningful as possible.

4. Look for the common language used in a particular kind of problem. Pay close attention to the word problems on the lesson practice pages, as they model the different kinds of language that may be used for the new concept just studied. For example, "altogether" often indicates addition. These "key words" can be useful clues, but they should not be a substitute for understanding.

5. Look for practical applications that use the concept and ask questions in that context.

6. Have the student invent word problems to illustrate number problems from the lesson.

Cautions:

1. Unneeded information may be included in the problem. For example, you may be told that Suzie is eight years old, but the eight is irrelevant when adding up the number of gifts she received.

2. Some problems may require more than one step to solve. Model these questions carefully.

3. There may be more than one way to solve some problems. Experience will help the student choose the easier or preferred method.

4. Estimation is a valuable tool for checking an answer. If an answer is unreasonable, it is possible that the wrong method was used to solve the problem.

LESSON 2

Place Value
with Expanded and Exponential Notation

In the base 10 (decimal) system, all numbers are represented by the digits 0 to 9 and by the place values of units, tens, hundreds, thousands, etc. There are several ways to take a number and break it down into these components to make our system of *decimal notation* more explicit. This lesson will explore three of them. The first way, which should be review, is *place-value notation*. In that notation, 156 can be shown as 100 + 50 + 6. The second way is *expanded notation*. In expanded notation, 156 can be renamed as 1 × 100 + 5 × 10 + 6 × 1, showing the individual digits multiplied by each place value. The third way is *exponential notation*. Using what you have learned about exponents, you can express all the place values in expanded notation as 10 to a power.

Figure 1 on the next page shows several of the place values, beginning with the units place and continuing through the ten thousands place. The first three places are easily shown with the blocks as in the figure, but the thousands place poses a challenge. One way to build 1,000 is a cube that is 10 by 10 by 10. (See Figure 2.) Following the progression where the exponent indicates the dimension, use the second power for two dimensions and the third power for three dimensions. You can't show 10,000 this way. This is because 10,000 would be in the fourth dimension, and you can't draw or construct a four-dimensional figure. Instead, you can show each of the place values two-dimensionally. The number 1,000 is shown as 10 by 100. The number 10,000 is 100 by 100. If you had enough space, you could even show 100,000 as 100 by 1,000 and 1,000,000 as 1,000 by 1,000.

Figure 1

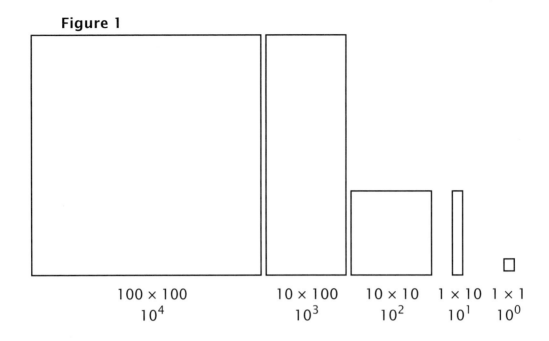

Figure 2

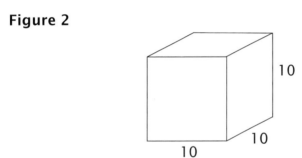

Notice how the place value corresponds to a power of 10 each time. The proof for $10^0 = 1$ will be shown in *Algebra 1*. Notice how the exponent for each power of 10 corresponds to the number of zeros in the place value. When you multiply by 10, another zero is added. 10^3 is 1,000, which has three zeros, and 10^1 is 10, which has one zero. In the examples following Figure 3, a number is shown first in expanded notation and then in exponential notation.

Figure 3

_____ , _____ _____ _____ , _____ _____ _____
1,000,000   100,000   10,000   1,000   100   10   1
10^6 10^5 10^4 10^3 10^2 10^1 10^0

Example 1

Express 245 in expanded notation and in exponential notation.

$$245 = 2 \times 100 + 4 \times 10 + 5 \times 1$$
$$245 = 2 \times 10^2 + 4 \times 10^1 + 5 \times 10^0$$

Example 2

Express 1,759 in expanded notation and in exponential notation.

$$1,759 = 1 \times 1,000 + 7 \times 100 + 5 \times 10 + 9 \times 1$$
$$1,759 = 1 \times 10^3 + 7 \times 10^2 + 5 \times 10^1 + 9 \times 10^0$$

Example 3

Express 803 in expanded notation and in exponential notation.

$$803 = 8 \times 100 + 3 \times 1$$
$$803 = 8 \times 10^2 + 3 \times 10^0$$

Example 4

Express $5 \times 1,000 + 6 \times 100 + 4 \times 10$ in decimal notation.

$$5 \times 1,000 + 6 \times 100 + 4 \times 10 = 5,640$$

LESSON 3

Decimal Numbers with Expanded Notation

Decimal numbers are fractions written in the base 10 system. They are a blend of place value and of fractions. This lesson will go back and forth between fractions and place value to show how decimals relate to all that has been learned up to this point. As well as learning how to add, subtract, multiply, and divide decimals, you should learn the concepts that will help you understand this subject.

Some teachers refer to decimals as ***decimal fractions.*** This is because decimals are fractions written in the same base 10, or decimal notation, which you have already been using for whole numbers. In Figure 1, a fraction is changed to an equivalent fraction with a denominator of 10. After constructing 2/5 with the overlays, place the clear overlay with 10 spaces on top of 2/5 to make tenths. The result is 4/10, or four tenths.

Figure 1

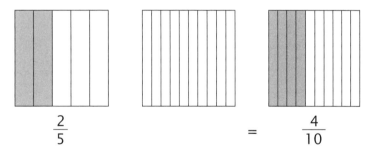

In expanded notation, 34 is shown as 3×10 + 4×1. The digits are 3 and 4, and the values are ten and one. A fraction may also be written in expanded notation. The fraction 2/5 can be written as 2 × ⅕. The digit 2 tells how many, and the value 1/5 tells what kind. In the decimal system, there is no fifths place. In the past, you have not dealt with any place value less than one (units). Now we need some smaller place values.

Recall that as you moved from right to left to increase the place values, you multiplied by a factor of 10. As you decrease and move from the greater to the lesser values and from left to right, you divide by a factor of 10.

Figure 2

$$\underline{} \swarrow \underline{} \swarrow \underline{} \swarrow \underline{} \swarrow \underline{} \swarrow \underline{}$$

100,000 / 10,000 / 1,000 / 100 / 10 / 1
10·10,000   10·1,000   10·100   10·10   10·1

100,000 ↘ 10,000 ↘ 1,000 ↘ 100 ↘ 10 ↘ 1
         100,000÷10   10,000÷10   1,000÷10   100÷10   10÷10

Since you are looking for values less than one, or fractional values, you need to continue this pattern of decreasing by, or dividing by, a factor of 10. From studying fractions, you know that dividing by 10 is the same as multiplying by 1/10, since the symbol "/10" means "divided by 10." Picking up where you left off, divide 1 by 10 or multiply 1 by 1/10.

Figure 3

1 ↘ $\frac{1}{10}$ ↘ $\frac{1}{100}$ ↘ $\frac{1}{1,000}$

1÷10   (1/10)÷10   (1/100)÷10
1×(1/10)   (1/10)×(1/10)   (1/100)×(1/10)

All you need now is a symbol to separate the whole-number place values from the fractional place values. As in money, the symbol is a decimal point. Between the units place and the tenths place, put a decimal point. See Figure 4 for the new fraction values written along with the numbers for units through hundreds. Notice that the units place, and not the decimal point, is the pivotal place.

Figure 4

$$\overline{}\ \overline{}\ \overset{\downarrow}{\overline{}}\ .\ \overline{\dfrac{1}{10}}\ \overline{\dfrac{1}{100}}$$

$$100\quad 10\quad 1$$

You now have values that are less than one unit. Refer back to Figure 2, where 2/5 was changed to 4/10. This was done because there is not a 1/5 place in the decimal system. However, there is a 1/10 place. In expanded notation, 4/10 is written as $4 \times \frac{1}{10}$. The fraction 4/10 may also be written as the decimal 0.4 (Figure 5). It can be expressed as 4 in the tenths place, or 4 tenths, or 0.4. The zero added to the left of the decimal point shows that there are no other values in the unit place. It calls attention to the unit place as well as the decimal point.

Figure 5

$$\overline{}\ \overline{}\ \overline{}\ .\ \overset{4}{\overline{\dfrac{1}{10}}}\ \overline{\dfrac{1}{100}}$$

$$100\quad 10\quad 1$$

Money is one place where decimal notation is commonly used to show fractions. If one unit is one dollar, 1/10 of a dollar is one dime, and 1/100 of a dollar (or 1/10 of a dime) is one cent, or one penny. Two fifths of a dollar is the same as 4/10 of a dollar or four dimes, as shown in Figure 5. There will be more about money on the worksheets and in another lesson.

Study the examples below to add to your understanding of decimals.

Example 1
Express 34.762 with expanded notation.
$$34.762 = 3 \times 10 + 4 \times 1 + 7 \times \dfrac{1}{10} + 6 \times \dfrac{1}{100} + 2 \times \dfrac{1}{1,000}$$

Example 2
Express 590.184 with expanded notation.
$$590.184 = 5 \times 100 + 9 \times 10 + 1 \times \dfrac{1}{10} + 8 \times \dfrac{1}{100} + 4 \times \dfrac{1}{1,000}$$

You can also write Examples 1 and 2 in exponential notation. Remember that 10^0 is used to represent the units place. You will learn why $10^0 = 1$ in *Algebra 1*.

Example 1 (in exponential notation)
$$34.762 = 3 \times 10^1 + 4 \times 10^0 + 7 \times \frac{1}{10^1} + 6 \times \frac{1}{10^2} + 2 \times \frac{1}{10^3}$$

Example 2 (in exponential notation)
$$590.184 = 5 \times 10^2 + 9 \times 10^1 + 1 \times \frac{1}{10^1} + 8 \times \frac{1}{10^2} + 4 \times \frac{1}{10^3}$$

LESSON 4

Add Decimal Numbers

In this lesson, you will begin using the algebra-decimal inserts to represent decimals. Turn a red hundred square upside down so the hollow side is showing and snap the flat green piece (from the algebra/decimal inserts) into the back. Then turn over several blue 10 bars and snap the flat blue pieces (also from the inserts) into their backs. Then take out the little one-half inch red cubes.

The large green square represents one unit. The size of the unit has been increased from the little green cube to this larger size, just as was done when you studied fractions. Since the large green square represents one, what do you think the flat blue bars represent? It takes ten of them to make one, so they are each 1/10, or 0.1. The red cubes represent 1/100, or 0.01.

Figure 1 shows how to represent 1.56 or $1 \times 1 + 5 \times \frac{1}{10} + 6 \times \frac{1}{100}$ with the decimal inserts.

Figure 1

As has been said before, decimal notation is used for money. Figure 1 represents money with the green unit as one dollar. A blue 1/10 bar represents one dime, or 1/10 of a dollar. A red cube represents one cent, or one penny, which is 1/100 of a dollar or 1/10 of a dime.

Figure 2

When you add up the change in your pocket, you add dollars to dollars, dimes to dimes, and cents to cents. Thinking of how you count money helps you understand why the key concept for adding and subtracting decimals is adding units to units, tenths to tenths, hundredths to hundredths, and thousandths to thousandths. This shouldn't seem strange because you have been adding like place values, regrouping as necessary, since *Beta*.

The easiest way to distinguish the values and make sure you are combining like values is by writing the problem vertically so the decimal point in one number is directly above (or below) the decimal point in the other number. Lining up these points ensures that your place values are also lined up. You may only add or subtract two numbers if they have the same value.

When using the inserts, it is clear that you can only add the green to the green, the blue to the blue, etc. When you don't have enough inserts for greater numbers, always line up the decimal points. The same skills are used for adding decimals and money as for adding any number. Remember that decimals are in base 10. You've just learned some new decimal values.

Example 1
Add: 1.56 + 1.23

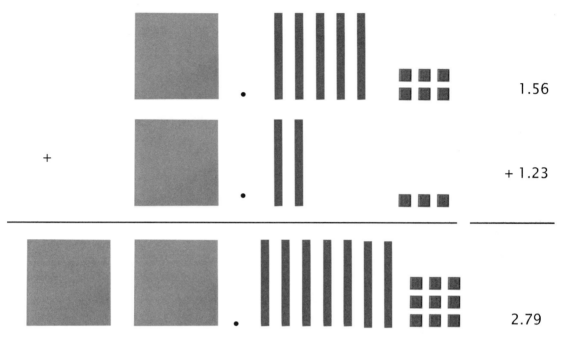

Example 2
Add: 1.79 + 0.54

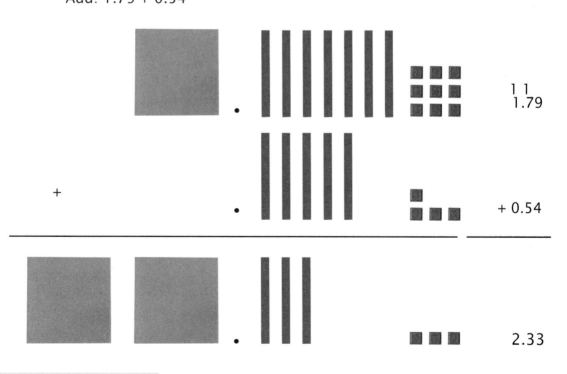

LESSON 5

Subtract Decimal Numbers

When subtracting decimals, the rules of regrouping still apply once you have lined up the place values. It helps to think of money when adding and subtracting decimals. In Examples 1 and 2, subtraction problems are illustrated. There are no blocks or inserts to represent subtraction, but at this stage the student should be able to subtract accurately and readily.

Example 1
Subtract: 1.76 − 1.42

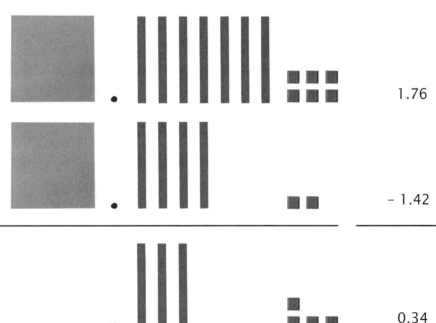

Example 2
Subtract: 2.29 - 1.75

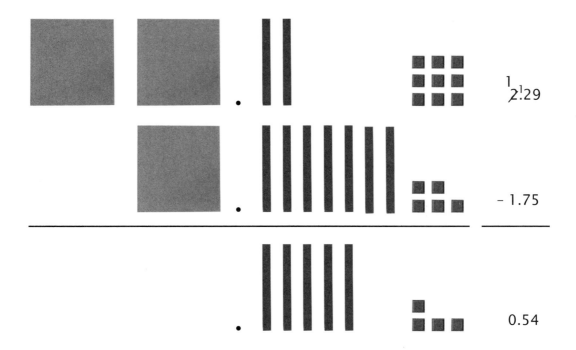

$\begin{array}{r} \overset{1}{2}\overset{!}{.}29 \\ -1.75 \\ \hline 0.54 \end{array}$

LESSON 6

Metric System Origin – Meter, Liter, Gram
and Greek Prefixes; Multi-step Word Problems

The term "metric" comes from the unit of length, the *meter*. ("Meter" comes from the word "metron," which refers to a device for measuring.) In the 1790s a group of French scientists were asked to create a consistent system of measurement. They decided to start with the distance from the North Pole to the Equator and to divide it into 10,000,000 equal pieces. Each of these pieces was defined as one "mètre" (in French). This is written as "meter" in American English. Today there are more accurate ways of measuring the earth, and a meter is defined as the distance light travels in a certain fraction of a second.

The metric system also uses different base units to measure things other than distance. The *gram*, for example, is used to measure mass (which is related to weight), and the *liter* is used to measure volume and capacity.

Rather than having several different units that measure the same thing (like using gallons, bushels, and teaspoons to measure volume), the metric system uses prefixes which can be put in front of any base unit to describe larger or smaller units. The three prefixes used in this lesson come from the Greek words for "thousand", "hundred", and "ten," respectively. Putting "kilo" before any base unit creates a unit that is equal to 1,000 of the base unit. For example, a kilogram is equal to 1,000 grams, and a kiloliter is equal to 1,000 liters. Each of the prefixes works the same way, and each of them (like each base unit) has an abbreviation.

Prefixes and Abbreviations
  kilo (k) = 1,000
  hecto (h) = 100
  deka (da) = 10

Base Units and Abbreviations

    meter (m)   indicates distance or length
    liter (l or L)   indicates volume and capacity
    gram (g)   indicates mass

You can combine these abbreviations in the same way that you combine the prefixes and base units. For example, you can combine the "k" (for kilo) and "m" (for meter) to get km (for kilometer). 1 km = 1,000 m

Be aware that in some places deka is spelled deca. Some older books use different abbreviations for deka, like "dk" or "D", but "da" is the official abbreviation. You should not use a lowercase "d" by itself because that stands for the prefix "deci," which you will be learning in the next lesson. Deka is the only metric prefix that has a two-letter abbreviation.

Liter can be abbreviated either with an uppercase "L" or a lowercase "l." The uppercase "L" is often used when the abbreviation is written by itself to avoid confusion with the numeral "1". You can use the lowercase "l" when it is combined with another abbreviation, such as "kl" for kiloliter.

As you learn these units, you can write the relationship between each prefixed unit and its base unit as an equation, as in 1 kilometer = 1,000 meters, or as a ratio:

$$\frac{1 \text{ kilometer}}{1{,}000 \text{ meters}}$$

You will learn more about ratios in enrichment pages 11G, 12G, 13G, and 14G. If you find it helpful, you could do some of these pages before returning to lesson 6. All of the ratios you will be writing in this lesson are equal to 1 because the amount on the top and the amount on the bottom are equal. Understanding this will be important in lesson 8 when you start converting between metric units.

Example 1

Fill in the spaces above and below the lines with the appropriate word, number, and abbreviation.

1. $\dfrac{1 \text{ kilogram (kg)}}{1{,}000 \text{ grams (g)}}$ $\dfrac{}{100 \text{ grams (g)}}$ $\dfrac{}{10 \text{ grams (g)}}$ $\dfrac{}{1 \text{ gram (g)}}$

2. _____ | 100 liters (L) | _____ | 1 liters (L)

3. _____ | _____ | 1 dekameter (dam) | 1 meter (m)
   1,000 meter (m)

Example 1 solution

1. $\dfrac{1 \text{ kilogram (kg)}}{1{,}000 \text{ grams (g)}}$ $\dfrac{1 \text{ hectogram (hg)}}{100 \text{ grams (g)}}$ $\dfrac{1 \text{ dekagram (dag)}}{10 \text{ grams (g)}}$ $\dfrac{1 \text{ gram (g)}}{1 \text{ gram (g)}}$

2. $\dfrac{1 \text{ kiloliter (kl)}}{1{,}000 \text{ liters (L)}}$ $\dfrac{1 \text{ hectoliter (hl)}}{100 \text{ liters (L)}}$ $\dfrac{1 \text{ dekaliter (dal)}}{10 \text{ liters (L)}}$ $\dfrac{1 \text{ liter (L)}}{1 \text{ liter (L)}}$

3. $\dfrac{1 \text{ kilometer (km)}}{1{,}000 \text{ meter (m)}}$ $\dfrac{1 \text{ hectometer (hm)}}{100 \text{ meter (m)}}$ $\dfrac{1 \text{ dekameter (dam)}}{10 \text{ meter (m)}}$ $\dfrac{1 \text{ meter (m)}}{1 \text{ meter (m)}}$

Multi-Step Word Problems

The student workbook includes some fairly simple two-step word problems. Some students may be ready for more challenging problems. Here are a few to try, along with some tips for solving this kind of problem. You may want to read and discuss these with your student as you work out the solutions together. The purpose is to stretch, not to frustrate. If you do not think the student is ready, you may want to come back to these later.

There are more multi-step word problems on the next page and in lessons 12, 18, and 24 of this instruction manual. The answers are at the end of the solutions at the back of this book.

1. Jill bought items that cost $3.45, $1.99, $6.59, and $12.98. She used a coupon worth $2.50. If Jill had $50.00 when she went into the store, how much money did she have when she left?

Although the problem asks only one question, there are other questions that must be answered first. The key to solving the problem is determining what the unstated questions are. Since the final question is asking for the leftover money, the unstated questions are: "What is the total of Jill's purchases?" and "What was the total bill after using the coupon?"

You might make a list of questions like this:

1. Total of purchases?

2. Total bill after subtracting coupon?

3. Leftover money?

2. Luke and Seth started out to visit Uncle Arnie. After driving 50 miles, they saw a restaurant, and Luke wanted to stop for lunch. Seth wanted to look for something better, so they drove on for eight miles before giving up and going back to the restaurant. After eating, they traveled on for 26 more miles from the restaurant. Seth saw a sign for a classic car museum, which they decided to visit. The museum was six miles from their route. After returning to the main road, they drove for another 40 miles and arrived at Uncle Arnie's house. How many miles is it from Luke and Seth's house to Uncle Arnie's house? How many miles did they drive on the way there?

The key to solving this is a careful drawing. It does not have to be to scale, but it should include all the parts of the journey.

3. Last week 3.5 inches of snow fell. Six tenths of an inch melted before another storm added 8.3 inches. Since then 4.2 inches of snow have melted. How many inches of snow are left on the ground?

LESSON 7

Metric System – Latin Prefixes

In the previous lesson you learned three prefixes that come from the Greek language and are used to create larger units of measure. In this lesson you will learn three prefixes which come from Latin and are used to create smaller units. *Milli* means one thousandth, while *centi* and *deci* mean one hundredth and one tenth respectively. You can abbreviate them with the letters m, c, and d.

$$\text{deci (d)} = \frac{1}{10}$$

$$\text{centi (c)} = \frac{1}{100}$$

$$\text{milli (m)} = \frac{1}{1,000}$$

One way to avoid confusing these with kilo, hecto, and deka is to notice that deci and centi have a soft "c" sound, like the "s" in "small." Another fact which may help is that American money uses a modified version of two of these prefixes. A decidollar (one tenth of a dollar) is called a dime. ("Dime" is the French form of the Latin decima.) A centidollar, or hundredth of a dollar, is called a cent. If you think "dime" when you see "deci," it may help you avoid confusing it with "deka."

Just as in lesson 6, these prefixes can be combined with the base units. The abbreviations may also be combined. 100 centimeters (cm) = 1 meter (m). The word "liter" by itself is abbreviated with a capital "L," but, when combined with a prefix, it may be represented by either an uppercase or a lowercase "l". Milliliters, for example, can be written as mL or as ml.

Example 1
Fill in the spaces above and below the lines with the appropriate word, number, and abbreviation.

1. $\dfrac{1 \text{ gram (g)}}{1 \text{ gram (g)}}$ $\dfrac{}{\frac{1}{10} \text{ gram (g)}}$ $\dfrac{}{}$ $\dfrac{}{}$

2. $\dfrac{}{}$ $\dfrac{}{\frac{1}{10} \text{ liter (L)}}$ $\dfrac{}{\frac{1}{100} \text{ liter (L)}}$ $\dfrac{1 \text{ milliliter (ml)}}{}$

3. $\dfrac{}{}$ $\dfrac{}{}$ $\dfrac{1 \text{ centimeter (cm)}}{}$ $\dfrac{}{\frac{1}{1{,}000} \text{ meter (m)}}$

Example 1 solution

1. $\dfrac{1 \text{ gram (g)}}{1 \text{ gram (g)}}$ $\dfrac{1 \text{ decigram (dg)}}{\frac{1}{10} \text{ gram (g)}}$ $\dfrac{1 \text{ centigram (cg)}}{\frac{1}{100} \text{ gram (g)}}$ $\dfrac{1 \text{ milligram (mg)}}{\frac{1}{1{,}000} \text{ gram (g)}}$

2. $\dfrac{1 \text{ liter (L)}}{1 \text{ liter (L)}}$ $\dfrac{\text{deciliter (dl)}}{\frac{1}{10} \text{ liter (L)}}$ $\dfrac{\text{centiliter (cl)}}{\frac{1}{100} \text{ liter (L)}}$ $\dfrac{1 \text{ milliliter (ml)}}{\frac{1}{1{,}000} \text{ liter (L)}}$

3. $\dfrac{1 \text{ meter (m)}}{1 \text{ meter (m)}}$ $\dfrac{1 \text{ decimeter (dm)}}{\frac{1}{10} \text{ meter (m)}}$ $\dfrac{1 \text{ centimeter (cm)}}{\frac{1}{100} \text{ meter (m)}}$ $\dfrac{1 \text{ millimeter (mm)}}{\frac{1}{1{,}000} \text{ meter (m)}}$

Including fractions in these ratios can make them difficult to use when solving problems. Here is a modified version. It will be used in lesson 8.

1 **kilo**unit	1 **hecto**unit	1 **deka**unit	1 unit	10 **deci**units	100 **centi**units	1,000 **milli**units
1,000 units	100 units	10 units	1 unit	1 unit	1 unit	1 unit

Comparing Metric Units

One liter is exactly one cubic decimeter or 1,000 cubic centimeters.

$1 \text{ L} = 1 \text{ dm}^3 = 1{,}000 \text{ cm}^3$

One cubic meter is exactly one kiloliter or 1,000 liters.

$1 \text{ m}^3 = 1 \text{ kl} = 1{,}000 \text{ L}$

One milliliter is exactly one cubic centimeter.

$1 \text{ ml} = 1 \text{ cm}^3$

One liter of water has a mass of approximately one kilogram, or 1,000 grams. One milliliter, or one cubic centimeter, of water has a mass of about one gram.

It may also help to note that the inside of a green Math-U-See unit block is close to one cubic centimeter or one milliliter, so one gram of water would come close to filling up one green unit block.

Comparing with Customary Units

You may wonder how some of the common metric units correspond to familiar U.S. customary units. Here are a few that are worth remembering.

One meter is a little over three feet (ft) or one yard (yd).

One meter is about 39.37 inches (in); one yard is 36 inches.

One centimeter is less than 1/2, or about 0.4 of an inch.

One inch is 2.54 cm, or approximately 2½ cm.

One liter is approximately 1.06 quarts (qt).

One ounce (oz) is approximately 28 grams.

One gram is about the mass of a small paper clip.

One kilogram is approximately 2.2 pounds (lb).

One kilometer is a little over 1/2 mile (mi), or approximately 0.6 of a mile.

LESSON 8

Metric System Conversion – Part 1

In the last two lessons you learned the meaning of these metric prefixes, as well as their abbreviations.

kilo (k) = 1,000

hecto (h) = 100

deka (da) = 10

deci (d) = $\frac{1}{10}$

centi (c) = $\frac{1}{100}$

milli (m) = $\frac{1}{1,000}$

You have also been writing these relationships as ratios in tables like these:

Figure 1

$\frac{1 \text{ kilounit}}{1,000 \text{ units}}$	$\frac{1 \text{ hectounit}}{100 \text{ units}}$	$\frac{1 \text{ dekaunit}}{10 \text{ units}}$	$\frac{1 \text{ unit}}{1 \text{ unit}}$	$\frac{1 \text{ deciunits}}{\frac{1}{10} \text{ unit}}$	$\frac{1 \text{ centiunits}}{\frac{1}{100} \text{ unit}}$	$\frac{1 \text{ milliunits}}{\frac{1}{1,000} \text{ unit}}$
$\frac{1 \text{ kilounit}}{1,000 \text{ units}}$	$\frac{1 \text{ hectounit}}{100 \text{ units}}$	$\frac{1 \text{ dekaunit}}{10 \text{ units}}$	$\frac{1 \text{ unit}}{1 \text{ unit}}$	$\frac{10 \text{ deciunits}}{1 \text{ unit}}$	$\frac{100 \text{ centiunits}}{1 \text{ unit}}$	$\frac{1,000 \text{ milliunits}}{1 \text{ unit}}$

In this lesson you will use these ratios as multiplication factors to convert larger units to smaller units. You will first learn the standard method, which involves multiplying by conversion ratios, and then look at an alternate method and a shortcut.

Method 1 (Standard Method)

The standard way of converting units using a ratio is to multiply the original unit by the appropriate ratio. For example, if you need to convert one kilogram to grams, you multiply by the ratio which shows the relationship between kilograms and grams. Because the amounts on the top and bottom of these ratios are the same, you can present each ratio either way without changing the value.

Figure 2

$$\frac{1 \text{ kilogram (kg)}}{1{,}000 \text{ grams (g)}} = \frac{1{,}000 \text{ grams (g)}}{1 \text{ kilogram (kg)}}$$

When choosing which ratio to use, you will need to arrange it so that the unit you are converting to is on top.

Example 1

$$1 \text{ kilogram (kg)} \times \frac{1{,}000 \text{ g}}{1 \text{ kilogram (kg)}} = 1{,}000 \text{ grams (g)}$$

Notice that you can cancel the unit kilograms because you are dividing kilograms by kilograms. After canceling, you will end up with an answer in grams. Also notice that when converting kilograms to grams you ended up multiplying by 1,000, which is exactly what "kilo" means.

It becomes slightly more complicated when you convert from one unit that has a prefix to another unit which has a prefix. For example, if you convert 1 kilometer to centimeters, it requires using two of the conversion ratios as factors.

Example 2

$$1 \text{ km} \times \frac{1{,}000 \text{ m}}{1 \text{ km}} \times \frac{100 \text{ cm}}{1 \text{ m}} = 1(1{,}000)(100 \text{ cm}) = 100{,}000 \text{ cm}$$

If you start with a number of units other than one, you just multiply that number by each factor along the way.

Example 3

$$15 \cancel{hl} \times \frac{100 \cancel{l}}{1 \cancel{hl}} \times \frac{10 \text{ dl}}{1 \cancel{l}} = 15(100)(10 \text{ dl}) = 15{,}000 \text{ dl}$$

If you need to convert between two units that are close to each other in size, you may end up both multiplying and dividing. For example, if you use the ratios to convert one centiliter to milliliters, it looks like this.

Example 4

$$1 \cancel{cl} \times \frac{1 \cancel{l}}{100 \cancel{cl}} \times \frac{1{,}000 \text{ ml}}{1 \cancel{l}} = 10 \text{ ml}$$

Method 2 (Alternative Method)

Think about the way conversions like this relate to place value. When you go from tens to hundreds, you multiply by 10, and when you go from hundreds to thousands, you multiply by 10 again. Because of this, if you create a ratio between any two units that are next to each other on our original table, you can describe them as a ratio of 10 to 1.

Figure 3

10 **hecto**units	10 **deka**units	10 units	10 **deci**units	10 **centi**units	10 **milli**units
1 **kilo**unit	1 **hecto**unit	1 **deka**unit	1 unit	1 **deci**unit	1 **centi**unit

If you are converting between two units that are adjacent to each other, these 10 to 1 ratios can be very helpful.

Example 5

$$3 \cancel{km} \times \frac{10 \text{ hm}}{1 \cancel{km}} = 30 \text{ hm}$$

$$2 \cancel{cm} \times \frac{10 \text{ mm}}{1 \cancel{cm}} = 20 \text{ mm}$$

If you are converting between units that are farther apart, you end up multiplying by the same ratio several times. You use the ratio once for each step that separated the units on the original chart. As shown in the video lesson, this is the same as multiplying by 10 for each step from larger units to smaller units on the chart.

Example 6

$$27 \,\cancel{hg} \times \frac{10 \,\cancel{dag}}{1 \,\cancel{hg}} \times \frac{10 \,\cancel{g}}{1 \,\cancel{dag}} \times \frac{10 \,dg}{1 \,\cancel{g}} = 27(10)(10)(10 \,dg) = 27{,}000 \,dg$$

Method 2 Shortcut

Each place in our decimal place value system is 10 times the place before it. Each metric unit is also 10 times the unit to the right of it on the chart. Because of this, we are, in effect, adding a zero for each step along the unit conversion chart as we go from larger to smaller units.

In summary, you have now learned two different ways for converting a larger metric unit to a smaller metric unit.

Example 7 (Method 1)

$$32 \,\cancel{hl} \times \frac{100 \,\cancel{l}}{1 \,\cancel{hl}} \times \frac{10 \,dl}{1 \,\cancel{l}} = 32(100)(10 \,dl) = 32{,}000 \,dl$$

Example 8 (Method 2)

$$32 \,\cancel{hg} \times \frac{10 \,\cancel{dag}}{1 \,\cancel{hg}} \times \frac{10 \,\cancel{g}}{1 \,\cancel{dag}} \times \frac{10 \,dg}{1 \,\cancel{g}} = 32(10)(10)(10 \,dg) = 32{,}000 \,dg$$

Because each step in the alternate method involves multiplying by 10 as you go from larger units to smaller units, you can use the shortcut of just changing the place value of the original number, thus "moving" the decimal place by adding a zero for each conversion step.

Example 9 (Method 2 Shortcut)

32. → 32000.

When doing the problems in the worksheets, you may use whichever method works best for you in a given problem.

LESSON 9

Multiply by 1/10 or 0.1

Multiplying decimals is based on what you already know of double-digit multiplication. Recognizing the dimensions and area of the inserts is important. Understanding the difference between the factors and the product is also critical. If needed, review the introduction to the decimal inserts in lesson 4.

The dimensions of the green unit are one over by one up, so the area is one. It represents a problem with the factors one by one and a product of one. The flat blue bar represents a product of 1/10 or 0.1, with an over factor of 1/10 and an up factor of one. Look at the figures below and think about this until you feel comfortable with the concept. Once you understand the blue bar, the red block will make more sense. It represents a product of 1/100 or 0.01, and the factors are 1/10 by 1/10, or 0.1×0.1. There is more explanation on the next page.

Figure 1

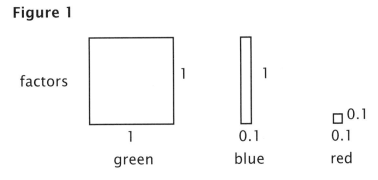

factors

green — 1 by 1
blue — 0.1 by 1
red — 0.1 by 0.1

Figure 2

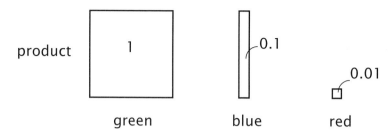

These figures look simple, but they are the key to understanding how to multiply decimals. Multiply the factors shown in Figure 1 to find the product in Figure 2. The unit block's factors are one and one, so one times one equals one. For the blue 1/10 bar, the factors are 1/10 and one, and 1/10 times one equals 1/10. Written with decimals, this is $0.1 \times 1 = 0.1$.

The red 1/100 piece has the factors 1/10 and 1/10. Using fractions, you are taking 1/10 of 1/10. In the three-step process for taking a fraction of a fraction, you divide 1/10 into ten equal parts and then take one of them. The result of 1/10 times 1/10 is 1/100. With decimals, it is $0.1 \times 0.1 = 0.01$. Money can help here. We know that 1/10 of a dime (which is 1/10 of a dollar) is a cent.

Example 1
Multiply: 1.2 times 0.4

Example 2
Multiply: 2.3 times 0.2

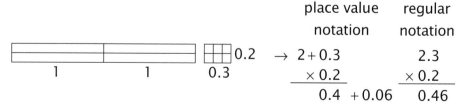

Example 3
Find the factors and area of this rectangle.

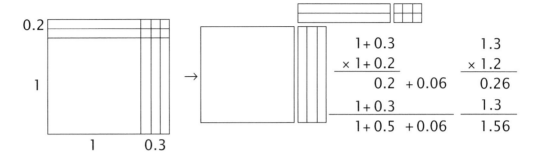

The factors are 1.3 by 1.2. The area is 1.56 square units.

Notice that when writing the problem, the up factor is on the line with the multiplication symbol, and the over factor is on the top line. Switching these factors will still produce the same answer, but it won't correspond with the picture shown above.

Math-U-See writes this differently than some so that the conceptual teaching comes first and the rules follow the concept. This is done for two reasons. First, most students expect a large answer when multiplying, as in $10 \times 20 = 200$. They usually see a product much greater than either factor. However, you are multiplying mixed numbers now, so your product is not going to be significantly greater than either factor. Second, this shows the reason for the rule normally employed when multiplying decimals, which is to multiply and then count the number of places to the right of the decimal point. Instead of moving the decimal point, you are going to do something "radical" and leave it stationary. Actually, moving decimal points is radical since they don't really move; it just appears that they do.

In the first step, 0.2×0.3 (or $2/10 \times 3/10$) shows that 0.2×0.3 is 0.06 (or $6/100$). Placing the 6 in the hundredths place shows two things. First, the product will be small, since you are starting two places to the right, and, second, this is where the "multiply and then count" comes from. Tenths (one place over) times tenths (one place over) yields hundredths in the answer. One place over plus one place over equals two places over. Then 0.2×1 is 0.2; 1×0.3 is 0.3; and 1×1 is 1. When adding, always add the same place values; $0.06 + 0$ is 0.06, $0.2 + 0.3$ is 0.5, and $1 + 0$ is 1. The answer is 1.56, as predicted.

Example 4

Find the factors and area of this rectangle.

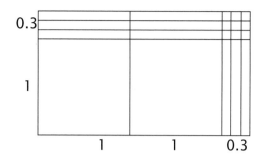

```
  2 + 0.3            2.3
× 1 + 0.3          × 1.3
  0.6  + 0.09       0.69
  2 + 0.3           2.3
  2 + 0.9 + 0.09    2.99
```

LESSON 10

Multiply Decimals by 1/100 or 0.01

This lesson will teach multiplying by hundredths. Emphasize thinking through each step and not merely memorizing. Even if you choose to multiply and then count as described later in the lesson, the thinking taught here is still invaluable for estimating the final answer. Begin by taking 1/100 or 0.01 of several numbers.

Example 1
Find 1/100 of 1. Remember that "of" in these problems means "times."

$$\frac{1}{100} \text{ of } 1 \text{ or } \frac{1}{100} \times \frac{1}{1} \text{ is } \frac{1}{100} \qquad 0.01 \times 1 = 0.01$$

Example 2
Find 1/100 of 100.

$$\frac{1}{100} \text{ of } 100 \text{ or } \frac{1}{100} \times \frac{100}{1} \text{ is } \frac{100}{100} \text{ or } 1 \qquad 0.01 \times 100 = 1$$

Example 3
Find 1/100 of 10.

$$\frac{1}{100} \text{ of } 10 \text{ or } \frac{1}{100} \times \frac{10}{1} \text{ is } \frac{10}{100} \text{ or } \frac{1}{10} \qquad 0.01 \times 10 = 0.1$$

Example 4
Find 1/100 of 1/10.

$$\frac{1}{100} \text{ of } \frac{1}{10} \text{ or } \frac{1}{100} \times \frac{1}{10} \text{ is } \frac{1}{1,000} \qquad 0.01 \times 0.1 = 0.001$$

Notice that when you multiply by 1/100, it is the same as dividing by 100. When you divide by 100, your answer moves two places to the right because it is that much less. It appears that the decimal point "moves" two places to the left. These are two ways of describing the same phenomenon. Multiplying by 1/10 (which is 0.1 in decimal form) or dividing by 10 moves the answer one place to the right, and the decimal point appears to move one place to the left.

Example 5

```
   2.3              23       one place
 × 0.32           × 32     + two places
  0.046    or      46
  0.69             69
  0.736           736
                 0.736      three places from the right
```

You know how to multiply; the key is place value. You can think, "hundredths times tenths is thousandths" and begin in the thousandths place with the partial products. Alternately, you can multiply and solve the problem as if there were no decimals and then count the number of spaces to the right of the decimal point in both factors (one space plus two spaces). Count from the right and put the decimal point after the third space from the right.

This is the same as thinking "hundredths times tenths is thousandths," multiplying as if there were no decimal points, and then placing the decimal point in the answer so that the last digit is in the thousandths place.

Example 6

```
  300.           300    zero places
× 0.02         ×   2  + two places
 6.00    or    600
```

               6.00   two places from the right

A hundredth times a hundred is one unit.

Example 7

```
   47.            47    zero places
×  0.33        ×  33  + two places
  1.41   or     141
 14.1           141
 15.51         1551
```

              15.51   two places from the right

A hundredth times a unit is one hundredth.

Example 8

```
   4.32          432    two places
×  0.65        ×  65  + two places
  0.2160  or   2160
  2.592        2592
  2.8080      28080
```

              2.8080   four places from the right

A hundredth times a hundredth is one ten-thousandth.

LESSON 11

Finding a Percentage

In a previous lesson you changed a fraction to a decimal by placing the tenth overlay on top of it. You can go a step further and place the other tenth overlay on top of the first one at a 90° angle. In Figure 1, 2/5 is changed to 4/10 and to 40/100. Notice that 4/10 = 0.4 and 40/100 = 0.40.

Figure 1

$$\frac{2}{5} = \frac{4}{10} = \frac{40}{100}$$

$$0.4 \qquad 0.40$$

Figure 2 shows how you might think of the change from a fraction to a percentage. The one and the two zeros from the number 100 were changed into a percent sign. This completes the change from the fraction (2/5) to a decimal (0.4) and then to a percentage (40%).

Figure 2

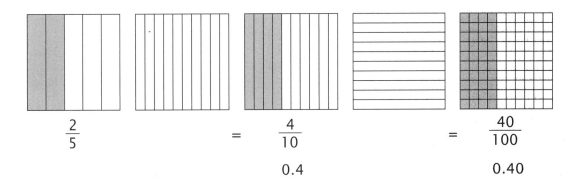

$$\frac{40}{100} \rightarrow \frac{40/}{00} \rightarrow \frac{40\%}{0} \rightarrow 40\% \rightarrow 40\%$$

Percent means "per hundred." A percentage is another way of writing hundredths. If you can change a fraction to hundredths (using either fraction or decimal notation), then you can easily change it to a percentage. The opposite is also true. If you have a percentage, you can immediately restate it as hundredths. To take a percentage of a number, simply change the percentage to hundredths and multiply it by the number.

Example 1
Find 25% of 36.

$25\% = 0.25$, and $0.25 \times 36 = 9$.
25% of 36 = 9

Another way to solve the same problem would be to change 25% to a fraction, simplify it, and multiply to find a fraction of a number.

Example 2
Find 25% of 36.

$$25\% = \frac{25}{100} = \frac{1}{4} \qquad \frac{1}{4} \times 36 = 9$$

Both ways are legitimate. Some problems are easier to solve using fractions, and some are easier with decimals.

Example 3
Find the tax on an order of onion rings that cost $1.50 if the tax rate is 8%.

$8\% = 0.08$, and $0.08 \times \$1.50 = \$0.12$

Example 4
Find the tip on a bill of $12.50 if you are tipping 16%.

$16\% = 0.16$, and $0.16 \times \$12.50 = \$2.00$

Example 5

Burgers are 20% off on Tuesdays. How much is a $2.80 burger on Tuesday?

20% = 0.20, and 0.20 × $2.80 = $0.56
$2.80 − $0.56 = $2.24

There is another way to solve Example 5. If you are going to be taking 20% off, the final answer will be 100% minus 20%, or 80% of the original amount. The total is 100%, or one. Eighty percent is 0.80, and 0.80 × $2.80 = $2.24. Instead of finding 20%, multiplying, and then subtracting, you are subtracting from 100% and then multiplying. Experiment by trying the same problem in different ways to see which method is more comfortable for you.

There are several percentages that may be helpful to memorize. They are listed below in Figure 3. You will be able to figure out some of the equivalents yourself when you learn to divide decimals later in this book. Since one fifth is 20%, then it follows that two fifths is double that, or 40%; three fifths is 60%; and four fifths is 80%.

Figure 3

$$\frac{1}{1} = 1.00 = 100\%$$

$$\frac{1}{2} = 0.50 = 50\%$$

$$\frac{1}{4} = 0.25 = 25\%$$

$$\frac{3}{4} = 0.75 = 75\%$$

$$\frac{1}{5} = 0.20 = 20\%$$

LESSON 12

Finding a Percentage > 100%
Word Problems

It is possible to have percentages greater than 100%. The percentage 100% is the same as one. Look at Figure 1. Percentages greater than 100% can be represented as improper fractions or as whole numbers.

Figure 1

$$1 = \frac{100}{100} = 100\% \quad 2 = \frac{200}{100} = 200\% \quad 3 = \frac{300}{100} = 300\%$$

Figure 2 shows percentages greater than one that can be written as **mixed numbers**.

Figure 2

$$1\frac{1}{2} = \frac{3}{2} = \frac{150}{100} = 150\% \text{ or}$$

$$1\frac{1}{2} = 1 + \frac{1}{2} = \frac{100}{100} + \frac{50}{100} = \frac{150}{100} = 150\%$$

$$2\frac{1}{4} = \frac{9}{4} = \frac{225}{100} = 225\% \text{ or}$$

$$2\frac{1}{4} = 2 + \frac{1}{4} = \frac{200}{100} + \frac{25}{100} = \frac{225}{100} = 225\%$$

In a problem that involves finding a tax, you can use the shortcut mentioned at the end of lesson 11 to save yourself some work. If onion rings are $1.50 with a tax rate of 8%, then the tax is 8%, or 0.08 × $1.50, which is $0.12. Adding $1.50 + $0.12 gives the actual cost of buying the onion rings, which is $1.62.

Instead of multiplying by 8% and then adding, you could have multiplied by 1.08. You have to pay the total cost of the onion rings, which is 100% of the cost, plus the tax, which is 8% of the cost. Therefore, 108% is the cost of the rings and the tax inclusive. Since 100% is the same as 1.00 and 8% is the same as 0.08, 108% is the same as 1.08. Multiply $1.50 times 1.08, and you get $1.62. Either method will work.

Example 1
Find the total cost for a meal priced at $12.50 if you are tipping 16% and the tax is 8%.

Total + tip + tax = 100% + 16% + 8% = 124%
124% × $12.50 = 1.24 × $12.50 = $15.50

Word Problems

Here are a few more multi-step word problems to try. You may want to read and discuss these with your student as you work out the solutions together. Again, the purpose is to stretch, not to frustrate. If you do not think the student is ready, you may want to come back to these later.

The answers are at the end of the solutions at the back of this book.

1. Naomi was ordering yarn from a website. If she ordered more than $50 worth of yarn labeled "discountable," she could take an extra 10% from the price of that yarn. Only some of the yarn she ordered was eligible for the discount. Here is what Naomi ordered:

    two skeins at $2.50 per skein   (not discountable)

    eight skeins at $8.40 per skein   (discountable)

    seven skeins at $5.99 per skein   (discountable)

    Tax is 5%, and shipping is 8%. How much did Naomi pay for her yarn?

2. After Naomi received her order, she found that she could make a scarf from one skein of the yarn that cost $5.99 a skein. After taking into account the discount, taxes, and shipping for the yarn, how much would it cost to make five scarves?

3. Naomi made a scarf for her brother using the yarn that cost $2.50 per skein. He offered to pay Naomi for the yarn she used. If she used 1.5 skeins of yarn to make the scarf, what was the total cost of the yarn?

LESSON 13

Reading Percentages in a Pie Graph

Circle graphs or *pie graphs* are often used to show percents. The whole pie is like the whole number one. If the whole pie is shaded, it represents 100%. If half of the pie is shaded, it represents one-half of 100%, or 50%. Each shaded part of the pie illustrates a fraction or a percentage of the whole. In the following examples, you will see data displayed visually with a pie graph.

Example 1

Colored Pictures Compared to Black and White Print

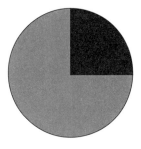

■ 25% of the book was illustrated with colored pictures.

▩ 75% of the book has black print on a white background.

Example 2

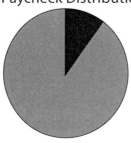
Paycheck Distributions

■ 10% of every paycheck is for giving.

■ 90% of the paycheck is for our living expenses.

Example 3

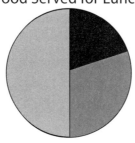
Food Served for Lunch

■ 20% of the noon meal was salad.

■ 30% of lunch was bread and rolls.

■ 50% of our lunch was soup.

Example 4

Handedness

■ 10% of the population are left-handed.

■ 85% of the people are right-handed.

■ 5% are ambidextrous.

LESSON 14

Multiply All Decimals

What do you think will happen when you multiply by 1/1,000, or 0.001? The answer will move three places to the right, or the decimal point will appear to move three places to the left. This is because you are taking 1/1,000 of the original number. In the examples below, the estimate is shown first. Then the problem is worked using both methods.

For the first method, line up the decimal points as in an addition problem. Then decide where to begin writing your answer. In Example 1, two hundredths times nine hundredths is 18 ten thousandths, so begin writing four spaces to the right of the decimal point.

For the second method, count the spaces to the right of the decimal point in both factors and add the spaces. Then multiply the numbers, paying attention only to the relative place value. Finally, count the spaces from the right, using the sum of the spaces in both factors, and place the decimal point.

Example 1

To estimate, round 75.89 to 80 and 0.52 to 0.5. Estimate: $80 \times 0.5 = 40$. Remember that 0.5 is the same as five tenths or one half, and one half of 80 is 40. Either way, the answer should be approximately 40.

```
      75.89              7589     two places
    × 0.52             ×   52   + two places
     1.5178     or      15178
    37.945              37945
    39.4628            394628
                        39.4628  four places from the right
```

Example 2

To estimate, round 536.042 to 500 and 0.08 to 0.1. Since 500 times 0.1 is 50, the answer should be approximately 50.

```
    536.042              536042    three places
  ×   0.08          ×         08  + two places
    42.88336  or       4288336
                       42.88336    five places from the right
```

Example 3

To estimate, round 703.091 to 700 and 1.584 to 2. Taking 700 times 2 gives 1,400. You can also think 1.584 is close to 1.5 or 1½, and 1½ times 700 is the same as 700 plus 350, or 1,050. Since 1.584 is between 1.5 and 2, the answer should be between our estimates of 1,050 and 1,400.

```
    703.091                703091     three places
  ×   1.584          ×        1584   + three places
      2.812364  or          2812364
     56.24728               5624728
    351.5455               3515455
    703.091                703091
  1,113.696144           1113696144
                         1,113.696144  six places from the right
```

LESSON 15

Metric System Conversions – Part 2

In Lesson 8, you learned two different methods for converting a larger unit to a smaller unit.

Example 1 (Standard Method)
Convert: 32 hectoliters = ____ deciliters

$$32 \, \cancel{hl} \times \frac{100 \, \cancel{l}}{1 \, \cancel{hl}} \times \frac{10 \, dl}{1 \, \cancel{l}} = 32(100)(10 \, dl) = 32{,}000 \, dl$$

Example 2 (Alternate Method)
Convert: 32 hectograms = ____ decigrams

$$32 \, \cancel{hg} \times \frac{10 \, \cancel{dag}}{1 \, \cancel{hg}} \times \frac{10 \, \cancel{g}}{1 \, \cancel{dag}} \times \frac{10 \, dg}{1 \, \cancel{g}} = 32(10)(10)(10 \, dg) = 32{,}000 \, dg$$

Because each step in the alternate method involved multiplying by 10, you were able to use the shortcut of "moving" the decimal place by one or by adding a zero for each conversion step.

In this lesson we will begin converting from smaller units to larger units. Watch how this works with both methods and with the shortcut.

Figure 1

1 **kilo**unit	1 **hecto**unit	1 **deka**unit	1 unit	10 **deci**units	100 **centi**units	1,000 **milli**units
1,000 unit	100 unit	10 unit	1 unit	1 unit	1 unit	1 unit

The standard method of multiplying by a conversion factor from this chart will not be affected by moving from smaller units to larger units. As before, you just need to be sure that, when you multiply by the conversion ratio, the top contains the unit you are converting to and the bottom contains the unit you are converting from. This way all the units will cancel out except the one you want in the end. Note that some of the numbers in this lesson will be in decimal form.

Example 3
Convert: _____ kilograms = 35 grams

$$35 \text{ g} \times \frac{1 \text{ kg}}{1{,}000 \text{ g}} = 0.035 \text{ kg}$$

<u>0.035</u> kilograms = 35 grams

Example 4
Convert: _____ kilometers = 670 centimeters

$$670 \text{ cm} \times \frac{1 \text{ m}}{100 \text{ cm}} \times \frac{1 \text{ km}}{1{,}000 \text{ m}} = 0.0067 \text{ km}$$

<u>0.0067</u> kilometers = 670 centimeters

The alternate method you used to convert larger units to smaller units was to multiply by a factor of 10 for each step, which moved the decimal point the same number of places to the left (or added the same number of zeros). When you multiply by a factor of 1/10, this is reversed, and the decimal point moves in the other direction.

Example 5
Convert: 3 deciliters = _____ milliliters

$$3 \text{ dl} \times \frac{10 \text{ cl}}{1 \text{ dl}} \times \frac{10 \text{ ml}}{1 \text{ cl}} = 3(10)(10) \text{ml} = 300 \text{ ml}$$

Shortcut: The decimal point moved 2 places to the right.
3.00 → 300.
3 deciliters = 300 milliliters.

Example 6
Convert: 280 centimeters = _____ hectometers

$$280 \text{ cm} \times \frac{1 \text{ dm}}{10 \text{ cm}} \times \frac{1 \text{ m}}{10 \text{ dm}} \times \frac{1 \text{ dam}}{10 \text{ m}} \times \frac{1 \text{ hm}}{10 \text{ dam}}$$

$$= 280 \left(\frac{1}{10}\right)\left(\frac{1}{10}\right)\left(\frac{1}{10}\right)\left(\frac{1}{10}\right) \text{ hm} = 0.028 \text{ hm}$$

Shortcut: The decimal point moved 4 places to the left.
00280. → 0.0280
280 centimeters = <u>0.028</u> hectometers

Notice that the decimal point appeared to move one place to the right for each step from larger units to smaller units (10/1 ratios) but appeared to move one place to the left for each step from smaller units to larger units (1/10 ratios). On the other hand, the decimal point was moving one place to the left for each step from larger to smaller and one place to the right for each step from smaller to larger.

Because it can be difficult when using the shortcut to remember whether the decimal point should be moving to the left or to the right, it can help to think of familiar situations from your experience, like changing money from smaller to larger units and vice versa.

It helps that American money uses a modified form of the same metric prefixes for tenths and hundredths, where instead of "decidollars" we have "dimes" and instead of "centidollars" we have "cents." As you know from experience, going from dollars to cents increases the number of individual units, but going from cents to dollars decreases the number of individual units.

Example 7
Convert: 5 dollars = _____ cents

Estimate: There should be more cents than dollars.

$$5 \text{ dollars} \times \frac{10 \text{ dimes}}{1 \text{ dollar}} \times \frac{10 \text{ cents}}{1 \text{ dime}} = 5(10)(10 \text{ cents}) = 500 \text{ cents}$$

Shortcut: The decimal moved 2 places to the right. 5.00 → 500
5 dollars = <u>500</u> cents

Example 8

Convert: ____ dollars = 700 cents

Estimate: There should be fewer dollars than cents.

$$700 \text{ cents} \times \frac{1 \text{ dime}}{10 \text{ cents}} \times \frac{1 \text{ dollar}}{10 \text{ dimes}}$$

$$= 700 \left(\frac{1}{10}\right)\left(\frac{1}{10}\right) \text{dollars} = 7 \text{ dollars}$$

Shortcut: The decimal point moved 2 places to the left.
700. → 7.00
<u>7</u> dollars = 700 cents

Example 9

Convert: ____ hectograms = 560 centigrams

Estimate: There should be more centigrams than hectograms.
The decimal point moved 5 places to the left. 000560. →
0.00560

<u>0.0056</u> hectograms = 560 centigrams

Example 10

Convert: 1.05 kilograms = ____ decigrams

Estimate: There should be fewer kilograms than decigrams.
The decimal point moved 4 places to the right. 001.05 → 10,500

1.05 kilograms = <u>10,500</u> decigrams

LESSON 16

Computing Area and Circumference
of a Circle with π ≈ 3.14

Computing the Circumference
Rectangles have area and perimeter, while circles have area and circumference. The *circumference* is the distance around the outside of a circle. The formula for the circumference of a circle is πd, or 2πr. The letter d stands for the *diameter*, the length of a line segment that passes through the center and touches both edges of a circle. ("Meter" means measure, and "dia" means through.)

Pi, or π, is the ratio of the circumference to the diameter. The Greek letter is used to represent this ratio because it is a number that cannot be written as a fraction. When the ratio is written as a decimal, it continues infinitely without repeating.

π = 3.14159265358979323846264338 . . .

For convenience, a number that is very close to π is often used instead. This lets you calculate a circumference which isn't exact but which can be written out as a fraction or as a decimal. One number that is very close to π is 22/7. Another is 3.14. You should use 3.14 as an approximation for π in this book because it gives you practice with multiplying decimals.

Since π is the ratio between the circumference of a circle and the diameter, the circumference by definition is equal to the diameter times π, which can be written as πd. When you know the diameter of a circle, you can use this formula to find the circumference, as in Example 1.

Example 1
Find the circumference of the circle.

Circumference = πd
Circumference ≈ (3.14)(10 in)
Circumference ≈ 31.4 in

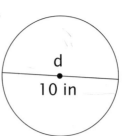

The *radius* is the distance from the center of the circle to the edge of the circle itself. To understand the common formula for the circumference, 2πr (where the letter r represents the radius of the circle), you can change the order of the three factors to π(2r). Since d (for diameter) is twice the radius, then πd = π(2r). The formula can be rearranged to 2πr, which is used in the next two examples.

Example 2
Find the circumference of the circle.

Circumference = 2πr
Circumference ≈ 2(3.14)(2.5 ft)
Circumference ≈ 15.7 ft

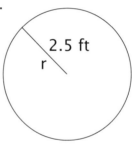

Example 3
Find the circumference of the circle.

Circumference = 2πr
Circumference ≈ 2(3.14)(8.4 ft)
Circumference ≈ 52.752 ft

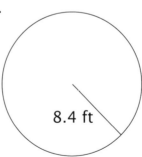

Computing the Area

The formula for the area of a circle is πr^2. The symbol π and the letter "r" have the same meanings as before, and the exponent "2" tells us that the radius is being used as a factor twice.

The following diagram may help you compare the area of a circle to the area of a square whose sides have the same length as the radius of the circle.

Figure 1

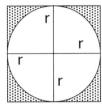

Notice that the number of squares with a side length of r and an area of r^2 which can combine to equal the area of the circle has to be some number between three and four. The actual number of squares that would cover the same area as the circle is π.

Example 4

Find the area of the circle.

Area = πr^2
Area ≈ 3.14(4.5 in)^2
Area ≈ 3.14(4.5 in)(4.5 in)
Area ≈ 3.14(4.5)(4.5)(in)(in)
Area ≈ 3.14(20.25)(in^2)
Area ≈ 63.585 in^2

Since these problems include measurement units, you are multiplying the units as well as the numbers. Think of 3 yd × 3 yd as the expression 3 × yd × 3 × yd. You can rewrite this as 3 × 3 × yd × yd = 9 yd^2. The exponent shows that you used the value of a yard as a factor two times.

Including units with your measurement calculations can help you see when units should be squared in your answer. For example, 3 cm × 4 cm = 12 cm^2 because you are multiplying the centimeters as well as the numbers. If you know the area of a rectangle and the length of one side and want to find the length of the other side, you would write it as 12 cm^2 ÷ 4 cm = 3 cm. Dividing cm^2 by cm gives just cm.

Example 5
Find the area of the circle.

Since the diameter is given, divide it by two to find the radius.

d = 2r; thus 13.6 ft = 2r, and 6.8 ft = r

Area = πr^2
Area ≈ $3.14(6.8 \text{ ft})^2$
Area ≈ $3.14(46.24 \text{ ft}^2)$
Area ≈ 145.1936 ft^2

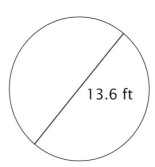

More on Perimeter and Area

The circumference of a circle corresponds to the perimeter of figures such as rectangles, squares, parallelograms, and triangles. Perimeter was taught in earlier levels of Math-U-See and is reviewed in the *Zeta* student book in the Quick Reviews for lessons 16, 17, and 18. Finding the area of figures other than circles was also taught in previous levels. Area is reviewed in lessons 19, 20, and 21 in the *Zeta* student book.

Introduction to Negative Numbers

Operations with **negative numbers** will be taught in detail in *Pre-Algebra*. However, students may already be encountering negative numbers in real-world contexts. Examples are temperatures above and below zero, elevations above and below sea level, positive and negative electrical charges, and credits and debits on a bank account.

Starting with lesson 15, several of the Application and Enrichment pages discuss negative numbers. It is also helpful to look for the applications mentioned above and discuss the meaning of the negative numbers involved. For example, –100 feet denotes a negative direction from sea level. However, if you dive to a depth of –100 feet, you will travel a distance of 100 feet, which is the absolute value of –100 feet.

The *absolute value* of a number is a positive value that describes distance rather than direction. It is indicated by two vertical lines on either side of a number.

$|4| = 4$, and $|-4| = 4$.

LESSON 17

Dividing a Decimal by a Whole Number

As for most math topics, there is "how to do it" and "why you do it." When dividing a decimal by a whole number, put the decimal point directly above the decimal point inside the "house" and then divide the numbers. The reason for this is related to multiplication. The missing factor on top of the house (known as the *quotient*), times the known factor (*divisor*) equals the product inside the house (*dividend*). Example 1 can be read as, "What number times 3 equals 1.2?" The answer, "0.4," is a decimal. We could also read it as "Three times what number equals 1.2?" We know that $0.4 + 0.4 + 0.4 = 1.2$, so $3 \times (0.4) = 1.2$. Another option is to focus on relative place value, as you have been doing with multiplication. The number 1.2 divided by 3 is 0.4, and 12 divided by 3 is 4. You can think about the problem as 12 divided by 3 and then make sure that the answer is written in the same place (tenths place) as the number you started with. See Example 1 below.

Figures 1 and 2 show how multiplication and division are related.

Figure 1

$$\text{factor} \overline{\smash{\big)}\ \text{product}}^{\text{factor}}$$

Figure 2

$$\text{divisor} \overline{\smash{\big)}\ \text{dividend}}^{\text{quotient}}$$

Example 1

$$3 \overline{\smash{\big)}\ 1.2} \rightarrow 3 \overline{\smash{\big)}\ 1.2}^{\ 0.} \rightarrow 3 \overline{\smash{\big)}\ 1.2}^{\ 0.4}$$

Always check your work: $3 \times (0.4) = 1.2$

Example 2

$$3\overline{)0.12} \rightarrow 3\overline{)\overset{0.}{0.12}} \rightarrow 3\overline{)\overset{0.04}{0.12}}$$

Always check your work: $3 \times (0.04) = 0.12$

In the examples, notice that the division itself isn't difficult. It is knowing where to put the places—or where to place the decimal point—that can be challenging. Put the decimal point directly above the decimal point in the dividend and then begin putting the missing factor directly above the number inside the "house." The four is directly above the two in both examples, and the decimal points are directly above each other. However, in Example 2, the "12" starts in the hundredths place; this means that "4" also goes in the hundredths place, and you put a zero in the tenths place.

Example 3
Estimate: Seven is close to eight, which is a multiple of four, so think, "What times four is equal to eight?" The answer is two, so seven divided by four is almost two.

```
      1.78        Check:
    _____           4
  4 )7.12         × 1.78                1.78
    -4             0.32        or      ×  4
     3 1           2.8                  7.12
    -2 8           4.
       32         _____
      -32          7.12
```

The fraction 1/8 is the same as 1 ÷ 8. The number 1 is the same as 1.0, or 1.000. You can add zeros to a number without changing its value as long as the zeros are not between the decimal point and a digit. The number 100.0 is not the same as 1.000. In Example 4, a fraction is changed into a decimal by dividing one by eight.

Example 4

Estimate: Eight is close to 10. Think, "What times 10 is equal to one?" The answer is one tenth or 0.1, so try 0.1 as the first step.

```
       0.125        Check:
    ┌───────           8
  8 │ 1.000         × 0.125              0.125
     -8              0.040      or        ×  8
     20              0.16                1.000
    -16              0.8
     40              1.000
    -40
```

Example 5

Estimate: 8.06 is close to 10. Think, "What times five is equal to 10?" The answer is two, so 8.06 divided by five is almost two.

```
       1.612        Check:
    ┌───────           5
  5 │ 8.060         × 1.612
     -5              0.010              1.612
     3 0             0.05       or      ×  5
    -3 0             3.0                8.060
      06             5.0
     -05             8.060
      10
     -10
```

In Example 5, you again needed to add a zero in order to finish solving the problem. Adding a zero at the end of the given digits *after* the decimal place is fine. If you added a zero in between the digits, such as between the 8 and the 6, the places of the digits would change, which would change the meaning of the numbers and lead to an incorrect answer.

Example 6

Estimate: 200.07 is close to 200, and 9 is almost 10, so think, "What times 10 is equal to 200?" The answer is 20, so try 2 as the first step.

```
        22.23
     ┌────────
   9 │ 200.07        Check:
      −18                 9
       20             × 22.23
      −18              0.27                22.23
       20              1.8       or        ×  9
      −18             18.                 200.07
        27           180.
       −27           ───────
                    200.07
```

LESSON 18

Dividing a Whole Number by a Decimal
and Word Problems

This lesson teaches a procedure for dividing a whole number by a decimal. Think it through using money for an example, and you will see why this procedure works.

In Example 1, the question is, "How many quarters of a dollar ($0.25), or how many groups of 25¢, are in $6.00?" Remember that $0.25 is a decimal, which is a another way to write a fraction. Since you are being asked how many fractions are in a whole number, the answer should be greater than the divisor.

Example 1

$$\$0.25 \overline{\smash{\big)}\ \$6}^{\ 24}$$

If there are four quarters in one dollar, then there must be four times six, or 24, quarters in six dollars.

The procedure for dividing a whole number by a decimal is to make the divisor (the number outside the box) a whole number so that you are dividing a number by a whole number. To make 0.25 a whole number, you need to multiply it by 100, and 100×0.25 is 25.

If you multiplied the divisor by 10, it would be 2.5, which is still a decimal. Recall that when you multiply the divisor by 100, you need to multiply the dividend by the same number as well; 100 times 6 is 600. See Example 2 on the next page.

Example 2

```
      24
25 ) 600
    -50
     100
    -100
```

$0.25 \times 100 = 25$, and $6 \times 100 = 600$

$$\frac{\$6}{\$0.25} \times \frac{100}{100} = \frac{600}{25}$$

Notice that this process is really the same as making an equivalent fraction. Multiplying the divisor (denominator) and dividend (numerator) by the same number doesn't change the quotient. When you multiply both numbers by a multiple of 10, whether it is 10, 100, or 1,000, the decimal points appear to move the same number of spaces in both numbers. No matter how many spaces you move the decimal point to make the divisor (on the outside) a whole number, you move the decimal point on the inside (the dividend) the same number of places.

Example 3

```
                            24.
0.25 ) 6.00  → 0.25 ) 6.00  → 25 ) 600
                           -50
                            100
                           -100
```

```
  0.25
× 24
  1.00
  5.0
  6.00
```

In Example 4, the question is "How many sets of three dimes (0.3 or 30¢) are in 18 dollars ($18.00)?" Remember that 0.3 and 0.30 are the same fraction written in different ways. Since you are asking how many fractions are in a whole number, the answer should be greater than 18.

Example 4

0.3 | 18

There are a little more than three sets of $0.30 in one dollar, so there must be at least three times 18, or 54 sets, in 18 dollars.

$$0.3\overline{)18} \rightarrow 0.3\overline{)18.0} \rightarrow 3.\overline{)180} \begin{array}{r} 60. \\ -18 \\ \hline 00 \\ -00 \end{array} \qquad \begin{array}{r} 60 \\ \times 0.3 \\ \hline 18.0 \end{array}$$

60 is greater than 54, so it fits with our estimate of 54.

In Example 5, the question is "How many sets of five hundredths of a dollar ($0.05 or 5¢) are in 121 dollars ($121.00)?" Remember that $0.05 is a fraction of a dollar.

Example 5

Since there are 20 nickels in a dollar, the answer should be 20 times greater than 121.

$$0.05\overline{)121} \rightarrow 0.05\overline{)121.00} \rightarrow 5.\overline{)12,100.} \begin{array}{r} 2,420. \\ -10 \\ \hline 2\ 1 \\ -2\ 0 \\ \hline 10 \\ -10 \\ \hline 00 \\ -0 \\ \hline 0 \end{array}$$

20 times 121 is 2,420. Our estimate was correct.

Word Problems

Here are a few more multi-step word problems to try. You may want to read and discuss these with your student as you work out the solutions together. Again, the purpose is to stretch, not to frustrate. If you do not think the student is ready, you may want to come back to these later.

The answers are at the end of the solutions at the back of this book.

1. The boys ordered several pizzas for the weekend. After the first evening, the following amounts of pizza were left over: one fourth of the pepperoni pizza, one half of the cheese pizza, three fourths of the mushroom pizza, and one fourth of the supreme pizza. The next morning, each boy ate the equivalent of one fourth of a pizza for breakfast. If that finished the pizza, how many boys were there?

2. Dan read that an average snowfall of 10 inches yields one inch of water when melted. He also found that five inches of very wet snow or 20 inches of very dry snow will both melt to an inch of water. He made measurements for a storm that started with 5.3 inches of average snowfall. The precipitation changed to wet snow and dropped another 4.1 inches. The weather continued to warm up, and the storm finished with 1.5 inches of rain. About how much water fell during the storm? Round your answer to tenths.

3. Jim bought edging to go around a circular garden with a radius of three feet. Later he decided to double the diameter of the garden. How many more feet of edging must he buy?

4. One packet of flower seeds was enough to just fill the area of the smaller garden in #3. How many packets of seed are needed for the larger garden? Round the area of each garden to the nearest square foot before solving the problem.

LESSON 19

Solving for an Unknown

An *algebraic expression* is a combination of one or more numbers, symbols representing numbers, and operation symbols. For example, X + 5 is a simple algebraic expression which has two *terms*. Just as words and phrases are combined to make a sentence, algebraic expressions may be combined to make equations. In previous levels of Math-U-See, you solved a number of different kinds of equations. In this book, you are going to solve equations that include decimals.

When solving for the unknown in an equation such as 0.3X = 12, you need to get the unknown (in this case, X) by itself. You want the final step to be X = ___ with a number in the blank. A *coefficient* means a factor within a term. Usually this word is used when discussing the number that indicates how many times a variable or unknown is being multiplied. In this case, 0.3 is the coefficient of X. To solve the equation, the coefficient must become 1. Think of the unknown as having no written coefficient because 1 times X is X.

Since 0.3X means X multiplied by 0.3, multiply by the reciprocal or use the inverse of multiplication, which is division, to make the coefficient of the unknown the number 1.

Example 1
Divide both sides by 0.3 so the left side will be 1 · X, or X without a written coefficient.

$$0.3X = 12$$
$$\frac{0.3X}{0.3} = \frac{12}{0.3} \rightarrow 0.3\overline{)12.} \rightarrow 3.\overline{)120.}^{40.}$$
$$X = 40$$

Check the answer.
$$0.3(40) = 12$$
$$12 = 12$$

Example 2

Divide both sides by 0.05 so the left side will be 1 · R, or R without a written coefficient.

$$0.05R = 6$$
$$\frac{0.05R}{0.05} = \frac{6}{0.05} \rightarrow 0.05\overline{\smash{)}6} \rightarrow 5.\overline{\smash{)}600.}^{\,120.}$$
$$R = 120$$

Check the answer.
$$0.05(120) = 6$$
$$6 = 6$$

LESSON 20

Dividing a Decimal by a Decimal

See if you can think through one of these problems using money as an example. The question is, "How many nickels are in 75 cents?"

Example 1

$$0.05 \overline{)0.75}^{\,15.}$$

There are 15 nickels in 75¢.

$15 \times (\$0.05) = \$0.75$

To solve these problems, first make the divisor, or the known factor, a whole number. To make 0.05 a whole number, multiply it by 100. The result is the whole number 5. Remember that you need to multiply *both* the divisor and the dividend by 100 to keep the values the same. Multiplying 0.75 times 100 gives 75. The original problem in new clothes is 75 ÷ 5. The answer in Example 2 is the same as in Example 1.

Example 2

```
      15
   _____
 5 | 75
    -5
    ___
     25
    -25
```

When you multiply both numbers by the same multiple of 10, the decimal points appear to move the same number of spaces in both numbers. In other words, no matter how many spaces the decimal point is moved to make the divisor (on the outside) a whole number, the decimal point on the inside (the dividend) must be moved the same number of places. In the following example, the number 3 is the same as 3.0.

Example 3

$$0.03 \overline{)1.974} \rightarrow 0.0\underset{\smile}{3} \overline{)1.\underset{\smile}{9}74} \rightarrow 3 \overline{)197.4}$$

$$\begin{array}{r} 65.8 \\ 3\overline{)197.4} \\ -18 \\ \hline 17 \\ -15 \\ \hline 2\,4 \\ -2\,4 \\ \hline \end{array}$$

$$\begin{array}{r} 0.0\,3 \\ \times\,65.8 \\ \hline 0.024 \\ 0.15 \\ 1.80 \\ \hline 1.974 \end{array} \qquad \begin{array}{r} 65.8 \\ \times\,0.0\,3 \\ \hline 1.974 \end{array}$$

Notice that in Example 4 you need to add a zero to the dividend because there aren't enough places to complete the division.

Example 4

$$0.006 \overline{)63.87} \rightarrow 0.00\underset{\smile}{6} \overline{)63.8\underset{\smile}{7}\underset{\smile}{} } \rightarrow 6 \overline{)63870}$$

$$\begin{array}{r} 10645 \\ 6\overline{)63870} \\ -6 \\ \hline 3 \\ -0 \\ \hline 38 \\ -36 \\ \hline 27 \\ -24 \\ \hline 30 \\ -30 \\ \hline \end{array}$$

$$\begin{array}{r} 10645 \\ \times\,0.006 \\ \hline 63.870 \end{array}$$

LESSON 21

Decimal Remainders

Remainders in a division problem involving decimals can be challenging. Here are four possibilities with accompanying examples.

Option 1

With certain single-digit divisors, such as 2, 4, 5, or 8, if you add enough zeros to the end of the dividend, eventually you will find the quotient without a remainder.

Example 1
Divide 2.3 by 8.

```
       0.2875
   8 ) 2.3000
      -16
        70
       -64
         60
        -56
          40
         -40
```

Option 2

With other single-digit divisors, such as 3, 6, 7, or 9, sometimes you can add all the zeros you want and never arrive at an answer without a remainder. It is with those problems in mind that we address options 2, 3, and 4. The most-used option is to select a place value that fits your needs and round to that value. For example, you may decide that you really don't need to know more than the hundredths place. Before you begin dividing in a case like this, decide the place to which you are going to round. This could be thousandths or millionths or any other place value, but, for Example 2, hundredths was chosen. All you need to do is divide to the thousandths place and then round to the closest value in the hundredths place.

Example 2
Divide 29.6 by 7.

```
            4.228
        7 ) 29.600
           -28
            16
           -14
             20
            -14
             60
            -56
```

4.228 rounds to 4.23.

Option 3

Another option for divisors such as 3, 6, 7, or 9 is to look for a pattern. Decimals with a pattern are called *repeating decimals.* When you divide by 3, eventually you will start seeing 3s in the quotient that go on and on. When you find a pattern like this, put a line above the repeating part to indicate that it goes on and on. Other divisors, like 7, have very long numbers in their patterns before they repeat, as in Example 3C. Study Examples A, B, and C.

Example 3A
Divide 8.5 by 3.
2.833 . . . equals 2.8$\overline{3}$.

```
           2.833
        3 ) 8.500
           -6
            25
           -24
            10
           - 9
            10
           - 9
```

Example 3B
Divide 4.6 by 9.

5.111 . . . equals $5.\overline{1}$.

```
       0.5111
    ┌─────────
  9 │ 4.6000
     -4 5
     ────
        10
        -9
        ──
        10
        -9
        ──
        10
        -9
```

Example 3C
Divide 9.5 by 7.

$\overline{1.3571428}$

The digits 571428 repeat.

Do you see where the pattern repeats again, beginning with 57?

```
       1.357142857
    ┌─────────────
  7 │ 9.500000000
     -7
     ──
      2 5
     -2 1
     ────
        40
       -35
       ───
         50
        -49
        ───
          10
          -7
          ──
           30
          -28
          ───
            20
           -14
           ───
             60
            -56
            ───
              40
             -35
             ───
               50
              -49
```

DECIMAL REMAINDERS - LESSON 21

Option 4

Occasionally it is necessary to express a decimal answer as a combination of a percentage and a fraction. For example, you might say that 33⅓ percent of the people surveyed liked a certain product. Example 4 was divided to the hundredths place, and the thousandths place written as a fraction. The 4 is written on top of the 7 because the next part of the problem is 4 divided by 7. The final answer is 0.27 4/7, or 27 4/7%.

Remember, though, that the 4/7 is really 0.04 divided by 7, so it represents 4/7 of a hundredth, not 4/7 of one. When you subtract 14 from 1.93, the first number is really 1.4. The 53 is really 0.53 if you leave in the decimal points. When you check the problem, 7 times 0.27 is 1.89, and 7 times 0.04/7 is 0.04. Then 0.04 is added to 1.89 for a total of 1.93. Think of the 4/7 as occupying the hundredths place along with the 7 in 0.27.

Example 4
Divide 1.93 by 7.

$$
\begin{array}{r}
27\frac{4}{7}\% \\
0.27\ \frac{4}{7} \\
7\overline{\smash{)}1.93} \\
\underline{-1\ 4} \\
53 \\
\underline{-49} \\
4
\end{array}
$$

LESSON 22

More Solving for an Unknown

Now you will learn how to solve more complex algebraic equations. In order to solve for the unknown X in 0.3X + 0.2 = 2, the unknown needs to be written by itself. The final step should be X = ___ with a number in the blank. You have learned how to make the coefficient 1. Now you will learn how to make the number added to 0.3X a zero. In the example, the number being added is 0.2.

The *variable,* or unknown, must be alone on one side of the equation, and everything else must be on the other side of the equation. The opposite, or inverse, of adding 0.2 is subtracting 0.2 from both sides.

Example 1

Subtract 0.2 from both sides so that the left side has just the variable.

Divide both sides by 0.3 so that the left side will be 1 · X, or just X.

$$0.3X + 0.2 = 2$$
$$\underline{-0.2 \quad -0.2}$$
$$0.3X + 0 = 1.8$$
$$0.3X = 1.8$$
$$\frac{0.3X}{0.3} = \frac{1.8}{0.3}$$
$$X = 6$$

$$\rightarrow 0.3 \overline{)18} \rightarrow 3. \overline{)18.}^{6.}$$

Check the answer.

$$0.3(6) + 0.2 = 2$$
$$1.8 + 0.2 = 2$$
$$2 = 2$$

Example 2

Subtract 3.3 from both sides so that the left side has just the variable.

$$1.5J + 3.3 = 13.8$$
$$-3.3 -3.3$$
$$1.5J + 0 = 10.5$$
$$1.5J = 10.5$$

Divide both sides by 1.5 so that the left side will be $1 \cdot J$, or just J.

$$\frac{1.5J}{1.5} = \frac{10.5}{1.5} \rightarrow 1.5\overline{)10.5} \rightarrow 15\overline{)105.}^{7.}$$
$$J = 7$$

Check the answer.

$$1.5(7) + 3.3 = 13.8$$
$$10.5 + 3.3 = 13.8$$
$$13.8 = 13.8$$

Example 3

$$0.12A + 0.052 = 0.1$$

Subtract 0.052 from both sides.

$$-0.052 -0.052$$
$$0.12A + 0 = 0.048$$
$$0.12A = 0.048$$

Divide both sides by 0.12.

$$\frac{0.12A}{0.12} = \frac{0.048}{0.12} \rightarrow 0.12\overline{)0.048} \rightarrow 12\overline{)4.8}^{0.4}$$
$$A = 0.4$$

Check the answer.

$$0.12(0.4) + 0.052 = 0.1$$
$$0.048 + 0.052 = 0.1$$
$$0.1 = 0.1$$

Volume

Finding the volume of cubes and other rectangular solids was taught in previous levels of Math-U-See. You can find review of these concepts in the *Zeta* student book in the Quick Reviews for lessons 22 and 23. A *rectangular prism* is another name for a rectangular solid.

LESSON 23

Convert Any Fraction
to a Decimal and a Percentage

Recall that the line separating a fraction into a numerator and denominator means "divided by." The fraction 1/4 means one divided by four. Now that you know how to divide numbers such as these, you can change any fraction into a decimal with hundredths. When dividing to make a percentage, always stop at the hundredths place because that is what percentages are—another way of writing hundredths. In Examples 2–4, make the answer stop at hundredths and write the thousandths place as a fraction. If you round the answer instead, it will be less precise. Once you have changed a number to hundredths, you can write it as a percentage. For more on final remainders, see lesson 21.

Example 1
Convert 1/4 to a decimal and a percentage.

$$\frac{1}{4} = 4\overline{\smash{)}1.00} \begin{array}{c} 0.25 \\ \underline{-8} \\ 20 \\ \underline{-20} \end{array} = 25\%$$

Example 2
Convert 2/3 to a decimal and a percentage.

$$\frac{2}{3} = 3\overline{\smash{)}2.00} \begin{array}{c} 0.66 \\ \underline{-1\,8} \\ 20 \\ \underline{-18} \\ 2 \end{array} = 0.\overline{6} \text{ or } 0.66\frac{2}{3} = 66\frac{2}{3}\%$$

CONVERT ANY FRACTION - LESSON 23

Example 3

Convert 3/7 to a decimal and a percentage.

$$\frac{3}{7} = 7\overline{)3.00}^{0.42} = 0.42\frac{6}{7} = 42\frac{6}{7}\%$$
$$\phantom{\frac{3}{7} = 7}\underline{-2\ 8}$$
$$\phantom{\frac{3}{7} = 7\ }20$$
$$\phantom{\frac{3}{7} = 7}\underline{-14}$$
$$\phantom{\frac{3}{7} = 7\ \ }6$$

Example 4

Convert 5/8 to a decimal and a percentage.

$$\frac{5}{8} = 8\overline{)5.00}^{0.62} = 0.62\frac{4}{8} = 0.62\frac{1}{2} = 62\frac{1}{2}\%$$
$$\phantom{\frac{5}{8} = 8}\underline{-4\ 8}$$
$$\phantom{\frac{5}{8} = 8\ }20$$
$$\phantom{\frac{5}{8} = 8}\underline{-16}$$
$$\phantom{\frac{5}{8} = 8\ \ }4$$

Example 5

Convert 1/3 to a decimal and a percentage.

$$\frac{1}{3} = 3\overline{)1.00}^{0.33} = 0.33\frac{1}{3} = 33\frac{1}{3}\%$$
$$\phantom{\frac{1}{3} = 3}\underline{-9}$$
$$\phantom{\frac{1}{3} = 3\ }10$$
$$\phantom{\frac{1}{3} = 3}\underline{-9}$$
$$\phantom{\frac{1}{3} = 3\ \ }1$$

Students should memorize the corresponding percentages for the fractions 1/2, 1/4, 3/4, and 1/5, if they have not done so already. Once one fifth is known, the other fifths may be easily found by multiplying by their numerators. For example, since one fifth is 20%, two fifths is 40%.

Figure 1 lists other fraction-decimal equivalents that you may want to memorize. They can be useful in everyday situations.

Figure 1

$$\frac{1}{3} = 33\frac{1}{3}\% \qquad \frac{1}{8} = 12\frac{1}{2}\%$$

$$\frac{2}{3} = 66\frac{2}{3}\% \qquad \frac{1}{9} = 11\frac{1}{9}\%$$

$$\frac{1}{7} = 14\frac{2}{7}\% \qquad \frac{1}{20} = 5\%$$

In many cases, these equivalents can be rounded if all you need is an approximate answer. You can think of one third as 33%, two thirds as 67%, one ninth as 11%, and so on. If you know one ninth is approximately 11%, then five ninths is five times that, or about 55%. The more you have tucked away in your memory, the more you will have readily available to solve problems. The more math you know, the more you can and will use.

A good application for transforming a fraction into a percentage is finding your percentage score when grading a test. First write a fraction showing how many answers are correct. If you get 18 right out of 20, your score is 18/20, or 90/100, or 90%.

LESSON 24

Decimals as Rational Numbers
and Word Problems

A *rational number* is a number that may be expressed as the result of dividing two numbers. A fraction, such as 7/8, is a rational number. Whole numbers may also be expressed as rational numbers, since 5 = 5/1. Decimals are fractions written in the base 10 system, so they are also a type of rational number. For example, to write the decimal 0.23 as a fraction, simply change it to 23/100. The numbers 0.23 and 23/100 are identical in value. One is written as a fraction, while the other is written as a decimal. For help in remembering the definition of a rational number, notice that the first five letters in R-A-T-I-O-nal refer to ratios.

Most of our study of ratios has been focused on fractions, but there were several application and enrichment pages (starting with 11G) that examined other types of ratios. For example, think of a classroom. If there are 10 boys and 15 girls, the ratio of boys to girls is 2 to 3, which can be written as 2/3. Since fractions show the ratio of the specific number of parts (numerator) to the number of those parts in a whole unit (denominator), fractions are a special type of ratio. Therefore, we can refer to fractions — and any other number that could be written as a fraction — as rational numbers. In the previous lesson, we converted fractions into decimals and into percentages. In this lesson, we will be doing the reverse: converting decimals into simplified fractions.

Example 1
Convert 0.29 into a simplified fraction.

$0.29 = \dfrac{29}{100}$ You can't simplify this fraction any further.

Example 2

Convert 0.8 to a simplified fraction.

$$0.8 = \frac{8 \div 2}{10 \div 2} = \frac{4}{5}$$

Example 3

Convert 0.036 to a simplified fraction.

$$0.036 = \frac{36 \div 4}{1000 \div 4} = \frac{9}{250}$$

Example 4

Convert 0.625 to a simplified fraction.

$$0.625 = \frac{625 \div 25}{1000 \div 25} = \frac{25 \div 5}{40 \div 5} = \frac{5}{8}$$

Example 5

Convert 0.035 to a simplified fraction.

$$0.035 = \frac{35 \div 5}{1000 \div 5} = \frac{7}{200}$$

Word Problems

Here are a few more multi-step word problems to try. You may want to read and discuss these with your student as you work out the solutions together. Remember that the purpose is to stretch, not to frustrate. If you do not think the student is ready, you may want to come back to these later.

The answers are at the end of the solutions at the back of this book.

1. Emily cut two circles from a sheet of colored paper measuring 8 inches by 12 inches. One circle had a radius of 3 inches, and the other had a radius of 2.5 inches. How many square inches of paper are left over? Is it possible to cut another circle with a 3-inch radius from the paper?

2. Tom wants to buy three items that cost $25.35, $50.69, and $85.96. He earns $6.50 an hour doing odd jobs. If 10 percent of his income is put aside for other purposes, how many hours must he work to earn the money for his purchases? Round your answer to the nearest whole hour.

3. Three tenths of the wooden toys were painted blue, and one fourth of them were painted green. Half of the remaining toys were painted red, and half were painted yellow. If 300 toys are blue, how many are there of each of the other colors?

LESSON 25

Mean, Median, and Mode

Statistics is the branch of mathematics that collects and organizes data to find patterns and make predictions or draw conclusions. Here are three terms used in statistics to describe the average or center of a set of data. Key letters in each word can help you remember them.

> meAn = A represents **A**verage; we usually are referring to the mean when we say "average."
>
> MeDian = MD represents the **M**iddle number in the **D**ata when arranged in ascending order.
>
> MOde = MO represents the number that occurs **M**ost **O**ften in the data.

Example 1
Arrange the following data in ascending order and find the mean, median, and mode: 5, 2, 5, 6, 8, 7, 9

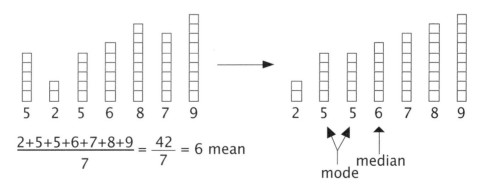

In Example 1, there is an odd number of data points, so the median is simply the middle point, which is six. When there is an even number of data, the median is calculated by finding the average of the two items in the middle. In Example 2, the median is the average of seven and nine, which is eight.

Example 2

Arrange the following data in ascending order and find the mean, median, and mode: 10, 2, 9, 4, 10, 7

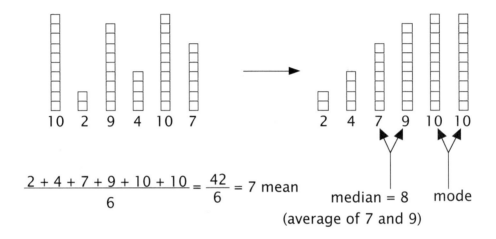

Statistics and Variability

Statistics is useful when a question has a number of different answers. "How tall are you?" has one correct answer. You don't need statistics to interpret the answer. On the other hand, "How tall are people in your family?" has a number of different answers. For the second question, the mean, median, and mode can be used to give you different ways to describe the center, or average, of the data.

Data can also be shown using a variety of charts and graphs. One example is a *dot plot*, which is a statistical chart that plots data on a simple number-line scale. Below is a dot plot showing the data from Example 2. Notice that the mode is easy to see from the chart but that you will still need to compute the mean and median in order to find the exact values.

LESSON 26

Probability

Probability is a numerical expression of how likely something is to happen or how likely a statement is to be true. It is expressed on a scale of 0 to 1, where a probability of 0 means that something is impossible and a probability of 1 means that it is certain. Probability is expressed as a ratio between a specific outcome and all possible outcomes:

$$\frac{\text{specified outcome}}{\text{possible outcomes}}$$

Example 1

You are rolling a six-sided die (singular of dice). On the sides are the numbers 1, 2, 3, 4, 5, and 6. Answer the following questions.

Question	Answer		
What is the probability of rolling a 4?	$\frac{\text{specified roll}}{\text{possible rolls}} =$	$\frac{4}{1-2-3-4-5-6}$	$= \frac{1}{6}$
What is the probability of rolling a 5?	$\frac{\text{specified roll}}{\text{possible rolls}} =$	$\frac{5}{1-2-3-4-5-6}$	$= \frac{1}{6}$
What is the probability of rolling a 1, 3, or 5 (an odd number)?	$\frac{\text{specified roll}}{\text{possible rolls}} =$	$\frac{1-3-5}{1-2-3-4-5-6}$	$= \frac{3}{6} = \frac{1}{2}$
What is the probability of rolling a 2, 4, or 6 (an even number)?	$\frac{\text{specified roll}}{\text{possible rolls}} =$	$\frac{2-4-6}{1-2-3-4-5-6}$	$= \frac{3}{6} = \frac{1}{2}$
What is the probability of rolling a 1 or a 5?	$\frac{\text{specified roll}}{\text{possible rolls}} =$	$\frac{1-5}{1-2-3-4-5-6}$	$= \frac{2}{6} = \frac{1}{3}$

Example 2

You bought a book with 24 pages in it. Of the 24, 18 had printing, four had pictures, and two pages were blank. Answer the following questions.

Question	Answer
What is the probability of opening to a picture?	$\dfrac{\text{specified pages}}{\text{possible pages}} = \dfrac{4}{24} = \dfrac{1}{6}$
What is the probability of opening to printing?	$\dfrac{\text{specified pages}}{\text{possible pages}} = \dfrac{18}{24} = \dfrac{3}{4}$
What is the probability of opening to a blank page?	$\dfrac{\text{specified pages}}{\text{possible pages}} = \dfrac{2}{24} = \dfrac{1}{12}$
What is the probability of opening to pictures or printing?	$\dfrac{\text{specified pages}}{\text{possible pages}} = \dfrac{4+18}{24} = \dfrac{22}{24} = \dfrac{11}{12}$

LESSON 27

Points, Lines, Rays, and Line Segments

Geometry is the study of shapes and their sizes and relative positions. In formal terms, it deals with spatial relationships. The word geometry comes from an ancient Greek term for surveying or measuring land. "Geo" means earth or land, and "metry" means measure. To measure land, it must be broken into smaller, more manageable pieces. The smallest possible piece is an imaginary figure, called a *point*. It has no measurable size, only position or location. Its width or length cannot be measured, so it has no dimensions (is zero-dimensional). A dot is drawn to show a point, which is actually too small to be seen. The dot is the "graph" of the point and represents the point. It is labeled with a capital letter. In Figure 1, the specific point is named "point A."

Figure 1

·A   point A

Infinity is a quantity that is greater than any number. It is represented by the symbol ∞. A line is a collection of points, side by side, that has infinite length and no width. It extends endlessly in two directions and (in the context of plane geometry) is perfectly straight. It is drawn with arrows on both ends. Because a line has only length, it is said to be one-dimensional.

Figures 2 and 3 show two ways to label lines. Figure 2 is labeled "line m." In Figure 3, two points (represented by capital letters) are chosen to name the line "line QR", or $\overleftrightarrow{RQ}$. Note that the line symbol on top of RQ has two arrows and that the order of the points does not matter. Any two points on a line may be used to name a line.

Figure 2

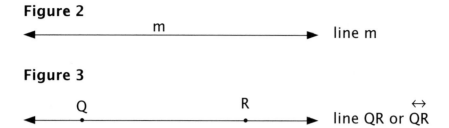

Figure 3

A ray may be thought of as part of a line. A *ray* is a geometric figure that has a specific starting point, called the endpoint or origin, at one end. The ray proceeds infinitely in only one direction. If you think of a laser pointer, the pointer itself is like the origin, and the laser beam goes on infinitely. Since a ray goes in only one direction, its symbol has only one arrow. Figure 4 is labeled as $\overrightarrow{BC}$ and read as "ray BC." The order the points are written is important with rays, as the first point listed is always the origin.

Figure 4

A *line segment* is a finite piece of a line with two endpoints. A line segment can be measured. Figure 5 is labeled as $\overline{LH}$ and read as "line segment LH." The same segment could be called $\overline{HL}$ because the order of the endpoints is not important. Only endpoints can be used to name a line segment.

Figure 5

Because points have no width or length, lines, rays, and line segments all contain an infinite number of points.

LESSON 28

Planes and Symbols

A *plane* is an infinite number of connected lines lying in the same flat surface. A plane has length and width, so it has two dimensions. You can think of it as an infinite floor or tabletop with no thickness. It is long and wide and keeps going in those two dimensions, but it is thinner than a piece of paper because it is just as thin as a line, which is as thin as a point. A picture of a parallelogram is used to represent a plane. A plane is often labeled with a lowercase letter. In Figure 1, the plane is referred to as "plane b."

Figure 1

This lesson and the previous one focus on flat, two-dimensional figures that lie in the same plane. This is called *plane geometry*. Three-dimensional geometry, with length, width, and height (or depth) pertaining to space and solids is called space or solid geometry. *Solid geometry* includes shapes like cubes, cylinders, pyramids, cones, and spheres.

To recap the lessons on geometry so far, consider that you operate in a three-dimensional place called space. The room you are in has the three dimensions of length, width, and height. Look at Figure 2 on the next page and notice how it progresses from no dimensions (a point) to one dimension (a line) to two dimensions (a plane) and finally to three dimensions.

Figure 2

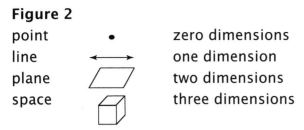

Symbols

The symbol "~" by itself means *similar,* or having the same shape but not exactly the same size. Two squares that have exactly the same shape but different measurements are said to be similar. Consider a square that is 2 inches by 2 inches and another square that is 4 miles by 4 miles. They have the same shape but are not the same size, so they are similar but not identical.

The equal sign "=" means exactly the same quantity, or *equal*, and is used if two numbers or quantities are the same. Putting the similar and equal signs together ($\cong$) means that two shapes have exactly the same proportions and that the dimensions of the shapes have the same numerical value. We say the shapes themselves are *congruent*. Use the equal sign for quantities (including measurable quantities like the length or area of a shape). Choose the congruent sign for objects that have the same shape and size.

Figure 3

similar	~	used to compare shapes
equal	=	used to compare numbers
congruent	$\cong$	used to compare shapes

LESSON 29

Angles

If two lines intersect, the opening, or space between the lines, is referred to as an *angle.* In Figure 1, there are four angles shown by the arcs. These angles are named 1, 2, 3, and 4.

Figure 1

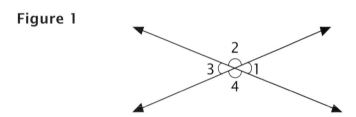

Figure 2 shows just one angle, made up of two rays with the same endpoint. This endpoint, or origin, is called the *vertex.* The plural of vertex is vertices.

Figure 2

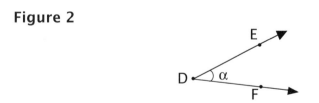

The rays are $\overrightarrow{DE}$ and $\overrightarrow{DF}$. The angle is labeled with either a number, as in Figure 1, or a lowercase Greek letter, as in Figure 2. The angle in Figure 2 is $\angle \alpha$ (angle alpha). Another way to identify this angle is by picking one point on each ray and the vertex to make either $\angle EDF$ or $\angle FDE$. Notice that the point labeling the vertex is always in the middle.

The size of angles is measured using degrees. Each degree is 1/360th of a full turn, or a complete circle.

A *right angle* has a measure of 90 degrees (90º) and forms a square corner. Figure 3 shows that the position of the angle doesn't matter; its measure is always 90º. Usually, a box symbol is used to indicate a right angle. Since there are 360 degrees in a circle, a right angle is one fourth of a circle, as shown in Figure 4.

Figure 3

Figure 4

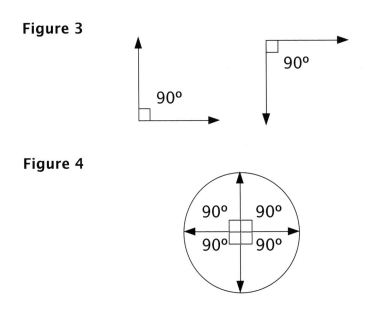

LESSON 30

Types of Angles

An *acute angle* is greater than 0° but less than 90°. Since it is a small angle, it may help to remember the name by thinking of "cute." Figure 1 has two acute angles.

Figure 1
0° < acute < 90°

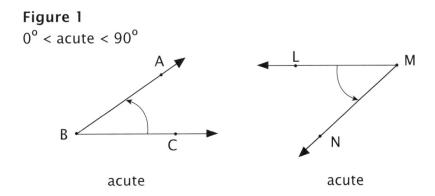

acute   acute

An *obtuse angle* is greater than 90° and less than 180°.

Figure 2
90° < obtuse < 180°

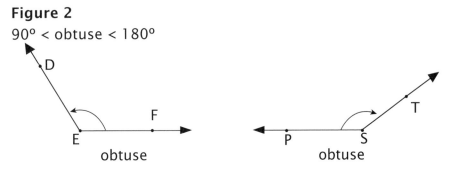

obtuse   obtuse

A *straight angle* is a less familiar angle. It has a measure of 180°.

Figure 3

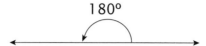

You sometimes hear of a car that skidded on ice and did a "one eighty," meaning it was going in one direction and then spun around until it was pointing in the opposite direction. This comes from the fact that it spun 180°. Figure 4 is a top view of this skid.

Figure 4

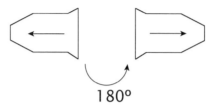

Student Solutions

Lesson Practice 1A
1. done
2. done
3. 4^2
4. 10^2
5. $5 \times 5 = 25$
6. $12 \times 12 = 144$
7. $4 \times 4 \times 4 = 64$
8. 6
9. $4 \times 4 = 16$
10. 100
11. done
12. 2^3
13. 5^2
14. 2^5
15. 8^2
16. 1^3
17. 4×4 or 16
18. 16 or 4×4

Lesson Practice 1B
1. 7^2
2. 12^2
3. 2^2
4. 6^2
5. $2 \times 2 \times 2 \times 2 = 16$
6. $1 \times 1 = 1$
7. $10 \times 10 = 100$
8. 8
9. $9 \times 9 = 81$
10. $3 \times 3 \times 3 = 27$
11. 2^1
12. 8^2
13. 2^4
14. 9^3
15. 6^4
16. 10^2
17. $5 \times 5 \times 5$ or 125
18. 125 or $5 \times 5 \times 5$

Lesson Practice 1C
1. 5^2
2. 8^2
3. 10^2
4. 1^2
5. $1 \times 1 \times 1 = 1$
6. $11 \times 11 = 121$
7. $2 \times 2 = 4$
8. $3 \times 3 \times 3 \times 3 = 81$
9. $8 \times 8 = 64$
10. $6 \times 6 = 36$
11. 7^2
12. 4^3
13. 9^2
14. 2^3
15. 7^5
16. 3^3
17. 10×10 or 100
18. 100 or 10×10

Lesson Practice 1D
1. $7 \times 7 = 49$
2. $12 \times 12 = 144$
3. $100 \times 100 = 10,000$
4. $1 \times 1 \times 1 \times 1 = 1$
5. 9
6. $3 \times 3 = 9$
7. 10^2
8. 5^3
9. 3^3
10. 4^2
11. 8^4
12. 5^3
13. $1 \times 1 \times 1$ or 1
14. 1 or $1 \times 1 \times 1$
15. done
16. $6 \div 3 = 2$
    $2 \times 1 = 2$

LESSON PRACTICE 1D - LESSON PRACTICE 2A

17. $20 \div 5 = 4$
    $4 \times 4 = 16$
18. $24 \div 6 = 4$
    $4 \times 5 = 20$
19. $\frac{1}{2}$ of 30 =
    $30 \div 2 = 15$
    $15 \times 1 = 15$ students
20. $\frac{7}{8}$ of 16 =
    $16 \div 8 = 2$
    $2 \times 7 = 14$ guests

Systematic Review 1E
1. $2 \times 2 \times 2 = 8$
2. $4 \times 4 \times 4 \times 4 = 256$
3. $11 \times 11 = 121$
4. $8 \times 8 = 64$
5. $6 \times 6 \times 6 = 216$
6. 15
7. 12^2
8. 6^2
9. 10^3
10. 1^5
11. 6^3
12. 9^2
13. 12×12 or 144
14. 144 or 12×12
15. $32 \div 8 = 4$
    $4 \times 3 = 12$
16. $12 \div 6 = 2$
    $2 \times 1 = 2$
17. $300 \div 3 = 100$
    $100 \times 2 = 200$
18. $72 \div 4 = 18$
    $18 \times 1 = 18$
19. $\frac{3}{4}$ of 600 =
    $600 \div 4 = 150$
    $150 \times 3 = 450$ visitors
20. $2 \times 12 = 24$ eggs
    $24 \div 6 = 4$
    $4 \times 1 = 4$ eggs

Systematic Review 1F
1. $9 \times 9 = 81$
2. $13 \times 13 = 169$
3. $3 \times 3 \times 3 \times 3 \times 3 = 243$
4. $5 \times 5 \times 5 \times 5 = 625$
5. 10
6. $20 \times 20 = 400$
7. 4^2
8. 7^3
9. 8^1
10. 3^4
11. 11^2
12. 10^3
13. $7 \times 7 \times 7 \times 7$ or 2,401
14. 2,401 or $7 \times 7 \times 7 \times 7$
15. $55 \div 5 = 11$
    $11 \times 2 = 22$
16. $210 \div 7 = 30$
    $30 \times 3 = 90$
17. $90 \div 10 = 9$
    $9 \times 1 = 9$
18. $54 \div 6 = 9$
    $9 \times 5 = 45$
19. 24 hours $\div 6 = 4$ hours
    4 hours $\times 1 = 4$ hours
20. 4×60 minutes = 240 minutes

Lesson Practice 2A
1. 10^3 ; 10^2 ; 10^1 ; 10^0
2. $10 \times 10 \times 10 \times 10 \times 10 \times 10 = 1,000,000$
3. 10
4. $10 \times 10 \times 10 \times 10 = 10,000$
5. $10 \times 10 \times 10 \times 10 \times 10 = 100,000$
6. 10^4
7. 10^6
8. 10^1

9. 10^2
10. done
11. $2\times100+7\times10+6\times1$;
    $2\times10^2+7\times10^1+6\times10^0$
12. $1\times1,000+4\times100+9\times1$;
    $1\times10^3+4\times10^2+9\times10^0$
13. $3\times10,000+1\times1,000+5\times100$;
    $3\times10^4+1\times10^3+5\times10^2$
14. 8,403
15. 70,060
16. 4,962
17. 3,530
18. 52,174

Lesson Practice 2B

1. 10^3; 10^2; 10^1; 10^0
2. $10\times10=100$
3. $10\times10\times10\times10\times10=100,000$
4. $10\times10\times10=1,000$
5. $10\times10\times10\times10\times10\times10=1,000,000$
6. 10^0
7. 10^2
8. 10^4
9. 10^3
10. $4\times1,000+8\times100+3\times10+6\times1$;
    $4\times10^3+8\times10^2+3\times10^1+6\times10^0$
11. $6\times100,000+2\times100+7\times10+5\times1$
    $6\times10^5+2\times10^2+7\times10^1+5\times10^0$
12. $3\times100+8\times10+4\times1$;
    $3\times10^2+8\times10^1+4\times10^0$
13. 5×10;
    5×10^1
14. 9,349
15. 617
16. 40,703
17. 2,874
18. 12,211

Lesson Practice 2C

1. 10^3; 10^2; 10^1; 10^0
2. $10\times10\times10\times10=10,000$
3. $10\times10=100$
4. $10\times10\times10\times10\times10\times10=1,000,000$
5. 1
6. 10^5
7. 10^1
8. 10^3
9. 10^6
10. $7\times100+2\times1$;
    $7\times10^2+2\times10^0$
11. $1\times10,000+1\times1,000+6\times100+8\times1$;
    $1\times10^4+1\times10^3+6\times10^2+8\times10^0$
12. $8\times1,000,000$;
    8×10^6
13. $4\times10+8\times1$;
    $4\times10^1+8\times10^0$
14. 5,607
15. 1,980
16. 770,000
17. 3,610,000
18. 216,534

Systematic Review 2D

1. 10,000
2. 100
3. 1,000
4. 0
5. 2
6. 4
7. $3\times1,000+7\times100+6\times10+6\times1$;
   $3\times10^3+7\times10^2+6\times10^1+6\times10^0$
8. $5\times10,000+1\times1,000+1\times10+7\times1$;
   $5\times10^4+1\times10^3+1\times10^1+7\times10^0$
9. 6,220
10. 10,901
11. $1^2=1$
12. $2^3=8$
13. $2^4=16$

SYSTEMATIC REVIEW 2D - LESSON PRACTICE 3A

14. $\frac{1}{3} = \frac{2}{6} = \frac{3}{9} = \frac{4}{12}$

15. $\frac{2}{5} = \frac{4}{10} = \frac{6}{15} = \frac{8}{20}$

16. $42 \div 7 = 6$;
    $6 \times 5 = 30$ ducks

17. 12 months $\div 4 = 3$ months
    The letter is J.

18. $24 \div 6 = 4$;
    $4 \times 5 = 20$ right-handed students
    $24 - 20 = 4$ left-handed students

Systematic Review 2E

1. 1
2. 1,000
3. 10
4. 10^1
5. 10^3
6. 10^0
7. $7 \times 100 + 4 \times 10 + 8 \times 1$;
   $7 \times 10^2 + 4 \times 10^1 + 8 \times 10^0$
8. $1 \times 10,000 + 2 \times 1,000 + 4 \times 100 + 6 \times 10 + 8 \times 1$;
   $1 \times 10^4 + 2 \times 10^3 + 4 \times 10^2 + 6 \times 10^1 + 8 \times 10^0$
9. 8,437
10. 60,294
11. $7^2 = 49$
12. $54^1 = 54$
13. $3^3 = 27$
14. $\frac{5}{6} = \frac{10}{12} = \frac{15}{18} = \frac{20}{24}$
15. $\frac{1}{4} = \frac{2}{8} = \frac{3}{12} = \frac{4}{16}$
16. $\frac{3}{8} = \frac{6}{16} = \frac{9}{24} = \frac{12}{32}$
17. $\frac{6}{7} = \frac{12}{14} = \frac{18}{21} = \frac{24}{28}$
18. 24 hours $\div 12 = 2$ hours;
    2 hours $\times 5 = 10$ hours
19. 12 months $\div 4 = 3$ months;
    3 months $\times 1 = 3$ months
20. 28 days $\div 4 = 7$ days;
    7 days $\times 3 = 21$ days

Systematic Review 2F

1. 100,000
2. 10,000
3. 100
4. 10^3
5. 10^4
6. 10^2
7. $5 \times 1,000 + 8 \times 100 + 8 \times 10 + 9 \times 1$;
   $5 \times 10^3 + 8 \times 10^2 + 8 \times 10^1 + 9 \times 10^0$
8. $6 \times 10,000 + 4 \times 100 + 1 \times 10$;
   $6 \times 10^4 + 4 \times 10^2 + 1 \times 10^1$
9. 7,260
10. 55,007
11. $9^2 = 81$
12. $1^5 = 1$
13. $2^4 = 16$
14. $\frac{2}{3} = \frac{4}{6} = \frac{6}{9} = \frac{8}{12}$
15. $\frac{3}{5} = \frac{6}{10} = \frac{9}{15} = \frac{12}{20}$
16. $\frac{1}{9} = \frac{2}{18} = \frac{3}{27} = \frac{4}{36}$
17. $\frac{7}{10} = \frac{14}{20} = \frac{21}{30} = \frac{28}{40}$
18. 60 minutes $\div 3 = 20$ minutes;
    20 minutes $\times 2 = 40$ minutes
19. $75 \div 5 = 15$;
    $15 \times 4 = 60$ bean plants
20. $30 \div 10 = 3$ packages damaged
    $30 - 3 = 27$ packages arrived safely

Lesson Practice 3A

1. 10; 1; $\frac{1}{100}$; $\frac{1}{1,000}$
2. multiply
3. divide
4. done
5. $6 \times 10 + 7 \times 1 + 2 \times \frac{1}{10} + 1 \times \frac{1}{100}$
6. done
7. $2 \times 10^0 + 7 \times \frac{1}{10^2}$
8. 1,643.119

9. 371.045
10. 68.95
11. done
12. dollar; dimes; cents; $1.00+$0.30+$0.08
13. dollar; dimes; cents; $1.00+$0.70+$0.02
14. dime
15. cent

8. 9,841.132
9. 3,006.084
10. 200.5
11. dollars; dimes; cents
    $7.00+$0.40+$0.05 = $7.45
12. 1; 1; 4; $1.00+$0.10+$0.04 = $1.14
13. dimes; cent; $0.60+$0.01 = $0.61
14. dollar
15. dollar

Lesson Practice 3B

1. 1,000; 100; 1; $\frac{1}{10}$; $\frac{1}{1,000}$
2. divide
3. multiply
4. $1 \times 1 + 9 \times \frac{1}{10} + 9 \times \frac{1}{100} + 9 \times \frac{1}{1,000}$
5. $2 \times 10 + 3 \times 1 + 6 \times \frac{1}{10} + 5 \times \frac{1}{100}$
6. $2 \times 10^2 + 3 \times 10^0 + 1 \times \frac{1}{10^2} + 6 \times \frac{1}{10^3}$
7. $8 \times 10^3 + 4 \times 10^2 + 3 \times 10^1 + 9 \times 10^0 + 7 \times \frac{1}{10^1}$
8. 2,401.613
9. 770.901
10. 5.008
11. dollars; dime; cents;
    $0.10; $0.09; $2.19
12. 3; 2; 7; $3.00+$0.20+$0.07 = $3.27
13. dimes; cents; $0.40+$0.05 = $0.45
14. cent
15. dime

Lesson Practice 3C

1. 1,000; 100; 10; $\frac{1}{10}$; $\frac{1}{100}$
2. left
3. right
4. $3 \times 100 + 4 \times 1 + 5 \times \frac{1}{100}$
5. $4 \times 1 + 6 \times \frac{1}{10} + 7 \times \frac{1}{100} + 9 \times \frac{1}{1,000}$
6. $6 \times 10^2 + 9 \times 10^1 + 1 \times 10^0 + 4 \times \frac{1}{10^1}$
7. $2 \times 10^1 + 5 \times 10^0 + 3 \times \frac{1}{10^1}$

Systematic Review 3D

1. $2 \times 10^3 + 3 \times 10^2 + 1 \times \frac{1}{10^1}$
2. $3 \times 10^1 + 8 \times 10^0 + 1 \times \frac{1}{10^1} + 2 \times \frac{1}{10^2} + 3 \times \frac{1}{10^3}$
3. 8,715.546
4. 6.411
5. dollars; dime; cents; $3.00; $0.09; $3.19
6. 9; 4; $9.00+$0.04 = $9.04
7. $5^2 = 25$
8. $1^5 = 1$
9. $10^3 = 1,000$
10. $10^0 = 1$
11. $\frac{1}{2} = \frac{2}{4} = \frac{3}{6} = \frac{4}{8}$
12. $\frac{5}{8} = \frac{10}{16} = \frac{15}{24} = \frac{20}{32}$
13. $\frac{8}{10} \div \frac{2}{2} = \frac{4}{5}$
14. $\frac{4}{24} \div \frac{4}{4} = \frac{1}{6}$
15. $\frac{6}{18} \div \frac{6}{6} = \frac{1}{3}$
16. $\frac{18}{30} \div \frac{6}{6} = \frac{3}{5}$
17. $4.00+$0.30+$0.06 = $4.36
18. 100¢ ÷ 10 = 10¢;
    10¢ × 4 = 40¢

Systematic Review 3E

1. $6 \times \frac{1}{10^1} + 1 \times \frac{1}{10^2} + 5 \times \frac{1}{10^3}$
2. $1 \times 10^2 + 3 \times 10^1 + 5 \times 10^0 + 4 \times \frac{1}{10^3}$

3. 451.221
4. 10.607
5. dollar; dimes; cents;
   $1.00 + $0.00 + $0.09 = $1.09
6. 6; 7; 5; $6.00 + $0.70 + $0.05 = $6.75
7. $3^2 = 9$
8. $4^3 = 64$
9. $2^2 = 4$
10. $10^3 = 1,000$
11. $\frac{4}{5} = \frac{8}{10} = \frac{12}{15} = \frac{16}{20}$
12. $\frac{3}{7} = \frac{6}{14} = \frac{9}{21} = \frac{12}{28}$
13. $\frac{11}{22} \div \frac{11}{11} = \frac{1}{2}$
14. $\frac{5}{25} \div \frac{5}{5} = \frac{1}{5}$
15. $\frac{4}{16} \div \frac{4}{4} = \frac{1}{4}$
16. $\frac{8}{32} \div \frac{8}{8} = \frac{1}{4}$
17. $1.00 + $0.40 + $0.07 = $1.47
18. 100¢ ÷ 5 = 20¢;
    20¢ × 3 = 60¢
19. $\frac{2}{4} = \frac{1}{2}$ of a melon
20. $10^2 = 100$ blocks

Systematic Review 3F

1. $1 \times 10^0 + 1 \times \frac{1}{10^3}$
2. $1 \times 10^3 + 3 \times 10^2 + 5 \times 10^1 + 8 \times 10^0 + 9 \times \frac{1}{10^1} + 1 \times \frac{1}{10^2}$
3. 6,528.05
4. 2,000.986
5. dollars; dimes; cents;
   $9.00 + $0.80 + $0.07 = $9.87
6. dollars; dimes; cents;
   $2.00 + $0.00 + $0.08 = $2.08
7. $3^4 = 81$
8. $1^3 = 1$
9. $10^0 = 1$

10. $5^2 = 25$
11. $\frac{9}{10} = \frac{18}{20} = \frac{27}{30} = \frac{36}{40}$
12. $\frac{1}{6} = \frac{2}{12} = \frac{3}{18} = \frac{4}{24}$
13. $\frac{5}{30} \div \frac{5}{5} = \frac{1}{6}$
14. $\frac{14}{35} \div \frac{7}{7} = \frac{2}{5}$
15. $\frac{20}{40} \div \frac{20}{20} = \frac{1}{2}$
16. $\frac{18}{27} \div \frac{9}{9} = \frac{2}{3}$
17. $6.00 + $0.09 = $6.09
18. 100¢ ÷ 100 = 1¢;
    1¢ × 7 = 7¢
19. $\frac{9}{12} = \frac{3}{4}$ of the dishes
20. 20 questions ÷ 5 = 4;
    4 × 4 = 16 questions

Lesson Practice 4A

1. done
2. done
3.   1.53
   +1.12
     2.65
4.   2.17
   +0.31
     2.48
5.   1.8
   +1.0
     2.8
6.   3.2
   +0.4
     3.6
7.   ¹
     1.13
   +1.68
     2.81
8.   ¹
     1.67
   +0.42
     2.09

LESSON PRACTICE 4A - LESSON PRACTICE 4B

9.   1.5
   +1.2
   ———
   2.7

10.   2.1
   +0.8
   ———
   2.9

11.   ¹
   1.16
  +1.46
  ———
   2.62

12.   3.90
  +0.02
  ———
   3.92

13.   ¹
   2.6
  +1.5
  ———
   4.1

14.   ¹
   1.8
  +1.3
  ———
   3.1

15.   3.00
  +1.62
  ———
   4.62

16.   4.48
  +0.10
  ———
   4.58

17. $4.51
  +$0.35
  ———
   $4.86

18.   ¹
   1.5
  +2.72
  ———
   4.22 miles

Lesson Practice 4B

1.   ¹
   7.1
  + 6.2
  ———
   13.3

2.   ¹
   5.9
  +1.2
  ———
   7.1

3.   ¹
   2.45
  +5.07
  ———
   7.52

4.   ¹
   4.13
  +1.96
  ———
   6.09

5.   7.0
  +2.8
  ———
   9.8

6.   1.5
  + 9.3
  ———
   10.8

7.   ¹
   8.84
  + 3.09
  ———
   11.93

8.   0.437
  +0.250
  ———
   0.687

9.   ¹
   8.8
  + 3.4
  ———
   12.2

10.   6.2
  +0.4
  ———
   6.6

11.   ¹
   2.70
  + 9.41
  ———
   12.11

12.   ¹
   5.52
  +0.60
  ———
   6.12

13.   3.9
  +4.0
  ———
   7.9

14.   ¹
   7.5
  +0.8
  ———
   8.3

15.   4.15
  +3.00
  ———
   7.15

LESSON PRACTICE 4B - SYSTEMATIC REVIEW 4D

16. 11
     0.524
     +0.277
     —————
     0.801

17. 1
     $$$12.95
     +$15.50
     —————
     $$$28.45

18. 1
     0.625
     +2.125
     —————
     2.750 gallons

10. 1
     2.8
     +5.9
     ————
     8.7

11. 1 1
     7.48
     +1.93
     ————
     9.41

12. 0.162
     +8.000
     —————
     8.162

13. 8.7
     +$$8.1
     ————
     16.8

14. 6.0
     +0.1
     ————
     6.1

15. 1
     0.731
     +$$0.402
     —————
     1.133

16. 1.125
     +0.112
     —————
     1.237

17. 4.3
     +0.5
     ————
     4.8 bushels

18. 2.045
     +0.5
     —————
     2.545 inches

Lesson Practice 4C

1. 1
    3.0
    +$$9.8
    ————
    12.8

2. 7.1
    +1.3
    ————
    8.4

3. 1 1
    1.95
    +$$8.15
    —————
    10.10

4. 1
    3.51
    +2.68
    ————
    6.19

5. 1
    5.9
    +0.4
    ————
    6.3

6. 4.1
    +3.0
    ————
    7.1

7. 1
    2.34
    +0.71
    —————
    3.05

8. 0.440
    +0.300
    —————
    0.740

9. 1
    6.5
    +$$5.0
    ————
    11.5

Systematic Review 4D

1. 1
    1.5
    +$$9.3
    ————
    10.8

2. 1
    5.9
    +1.6
    ————
    7.5

3. 6.34
    +2.41
    —————
    8.75

Systematic Review 4D

4. $\overset{1}{}1.82$
    $+\ 9.3$
    $\overline{11.12}$
5. $2^3 = 8$
6. $6^2 = 36$
7. $10^4 = 10,000$
8. $7^2 = 49$
9. $1 \times 100 + 7 \times 10 + 6 \times 1 + 2 \times \frac{1}{10} + 1 \times \frac{1}{100}$
10. $6 \times \frac{1}{10} + 8 \times \frac{1}{100} + 5 \times \frac{1}{1,000}$
11. $4 \times 1 + 5 \times \frac{1}{10}$
12. $\frac{1}{4} = \frac{2}{8} = \frac{3}{12} = \frac{4}{16}$
13. $\frac{5}{8} = \frac{10}{16} = \frac{15}{24} = \frac{20}{32}$
14. $\frac{1}{4} + \frac{3}{5} = \frac{5}{20} + \frac{12}{20} = \frac{17}{20}$
15. $\frac{3}{4} + \frac{1}{6} = \frac{18}{24} + \frac{4}{24} = \frac{22}{24} = \frac{11}{12}$
16. $\frac{1}{3} + \frac{2}{5} = \frac{5}{15} + \frac{6}{15} = \frac{11}{15}$
17. $\overset{1\ 1}{}75.25$
    $+\ \ 1.75$
    $\overline{77.00}$ inches
18. $12 \div 6 = 2$ spoiled apples
    $12 - 2 = 10$ good apples

Systematic Review 4E

1. $\overset{1\ 1}{}8.6$
    $+\ 2.4$
    $\overline{11.0}$
2. 3.0
    $+4.4$
    $\overline{7.4}$
3. $\overset{1\ \ 1}{}3.07$
    $+\ 9.25$
    $\overline{12.32}$
4. 5.00
    $+3.24$
    $\overline{8.24}$

Systematic Review 4F (from 4D column)

5. $3^4 = 81$
6. $5^2 = 25$
7. $1^7 = 1$
8. $10^3 = 1,000$
9. $4 \times 10^1 + 3 \times 10^0 + 3 \times \frac{1}{10^1}$
10. $6 \times 10^0 + 1 \times \frac{1}{10^1} + 5 \times \frac{1}{10^3}$
11. $2 \times 10^2 + 3 \times \frac{1}{10^1} + 4 \times \frac{1}{10^2}$
12. $\frac{1}{2} = \frac{2}{4} = \frac{3}{6} = \frac{4}{8}$
13. $\frac{9}{10} = \frac{18}{20} = \frac{27}{30} = \frac{36}{40}$
14. $\frac{1}{9} + \frac{1}{2} = \frac{2}{18} + \frac{9}{18} = \frac{11}{18}$
15. $\frac{2}{5} + \frac{5}{6} = \frac{12}{30} + \frac{25}{30} = \frac{37}{30} = 1\frac{7}{30}$
16. $\frac{1}{10} + \frac{2}{3} = \frac{3}{30} + \frac{20}{30} = \frac{23}{30}$
17. 0.5
    $+0.25$
    $\overline{0.75}$ hours
18. $\overset{1\ 1}{}9.5$
    $+11.6$
    $\overline{21.1}$ gallons
19. $\frac{2}{3} + \frac{1}{5} = \frac{10}{15} + \frac{3}{15} = \frac{13}{15}$ of the problems
20. $30 \div 15 = 2$;
    $2 \times 13 = 26$ problems

Systematic Review 4F

1. 5.6
    $+4.3$
    $\overline{9.9}$
2. $\overset{11}{}1.9$
    $+\ 9.2$
    $\overline{11.1}$
3. $\overset{1}{}5.13$
    $+\ 9.50$
    $\overline{14.63}$
4. $\overset{1\ 1}{}4.17$
    $+1.95$
    $\overline{6.12}$
5. $8^2 = 64$
6. $10^0 = 1$
7. $4^3 = 64$
8. $9^2 = 81$
9. $9,500.1$
10. 158.004

11. $\frac{1}{3} = \frac{2}{6} = \frac{3}{9} = \frac{4}{12}$

12. $\frac{3}{7} = \frac{6}{14} = \frac{9}{21} = \frac{12}{28}$

13. $\frac{2}{7} + \frac{1}{8} = \frac{16}{56} + \frac{7}{56} = \frac{23}{56}$

14. $\frac{3}{5} + \frac{2}{9} = \frac{27}{45} + \frac{10}{45} = \frac{37}{45}$

15. $\frac{3}{4} + \frac{1}{5} = \frac{15}{20} + \frac{4}{20} = \frac{19}{20}$

16. $2.25 + $1.69 = $3.94

17. $4.00 + $2.50 + $8.35 = $14.85

18. $\frac{5}{15} = \frac{1}{3}$

19. $\frac{3}{8} + \frac{1}{3} = \frac{9}{24} + \frac{8}{24} = \frac{17}{24}$; no

20. $27 \div 9 = 3$; $3 \times 5 = 15$ players

Lesson Practice 5A

1. done

2. 18.7 − 4.8 = 13.9

3. 9.35 − 8.46 = 0.89

4. 22.0 − 9.6 = 12.4

5. 6.4 − 5.3 = 1.1

6. 5.0 − 2.4 = 2.6

7. 23.60 − 9.43 = 14.17

8. 2.81 − 0.63 = 2.18

9. 4.0 − 2.6 = 1.4

10. 9.6 − .9 = 8.7

11. 8.93 − 5.00 = 3.93

12. 4.70 − 1.98 = 2.72

13. $7.17 − $2.98 = $4.19

14. 5.75 − 1.50 = 4.25 feet

15. 150.0 − 4.5 = 145.5 pounds

16. 15.15 − 4.29 = 10.86 seconds

17. $\overset{9}{\cancel{10}}.^{1}2$
     $-\ 0.7$
     $\overline{9.5 \text{ miles}}$

18. $\$\overset{9}{\cancel{10}}.\overset{14}{\cancel{5}}\overset{}{\cancel{0}}$
     $-\$\ 4.7\ 9$
     $\overline{\$\ 5.7\ 1}$

Lesson Practice 5B

1. 8.3
    -0.3
    $\overline{8.0}$

2. $\overset{8}{\cancel{9}}.^{1}1$
    -6.3
    $\overline{2.8}$

3. $\overset{4}{\cancel{5}}.\overset{9}{\cancel{0}}0$
    $-2.3\ 3$
    $\overline{2.6\ 7}$

4. 6.98
    -1.90
    $\overline{5.08}$

5. $\overset{4}{\cancel{5}}.^{1}1$
    $-\ 4.8$
    $\overline{0.3}$

6. 9.2
    -1.0
    $\overline{8.2}$

7. $8.\overset{5}{\cancel{6}}\overset{}{\cancel{0}}$
    $-5.3\ 5$
    $\overline{3.2\ 5}$

8. $12.\overset{6}{\cancel{7}}5$
    $-\ 9.0\ 6$
    $\overline{3.6\ 9}$

9. 1.3
    -0.6
    $\overline{0.7}$

10. $\overset{5}{\cancel{6}}.^{1}2$
    -1.8
    $\overline{4.4}$

11. $0.5\overset{4}{\cancel{9}}\overset{10}{\cancel{3}}$
    $-0.0\ 2\ 6$
    $\overline{0.4\ 8\ 7}$

12. 0.362
    -0.100
    $\overline{0.262}$

13. $\$\overset{010}{\cancel{1}\cancel{1}}.\overset{9}{\cancel{0}}0$
    $-\$\ 5.7\ 5$
    $\overline{\$\ 5.2\ 5}$

14. $\$3.\overset{4}{\cancel{5}}0$
    $-\$2.4\ 5$
    $\overline{\$1.0\ 5}$

15. $\overset{7}{\cancel{8}}.^{1}5$
    -7.6
    $\overline{0.9 \text{ hours}}$

16. $\overset{10}{\cancel{11}}.^{1}0$
    $-\ 4.5$
    $\overline{6.5 \text{ miles}}$

17. $12.4\overset{3}{\cancel{0}}\overset{9}{\cancel{0}}$
    $-\ 2.3\ 7\ 5$
    $\overline{10.0\ 2\ 5 \text{ years}}$

18. $\overset{1}{\cancel{2}}.\overset{10}{\cancel{1}}0$
    $-0.1\ 6$
    $\overline{1.9\ 4 \text{ inches}}$

Lesson Practice 5C

1. 7.8
    -1.2
    $\overline{6.6}$

LESSON PRACTICE 5C - SYSTEMATIC REVIEW 5D

2. $\begin{array}{r} \overset{3}{\cancel{4}}.\overset{1}{1} \\ -2.9 \\ \hline 1.2 \end{array}$

3. $\begin{array}{r} \overset{9}{\cancel{1}}0.\overset{9}{\cancel{0}}{}^{1}0 \\ -2.67 \\ \hline 7.33 \end{array}$

4. $\begin{array}{r} \overset{4}{\cancel{5}}.\overset{9}{\cancel{0}}{}^{1}1 \\ -1.92 \\ \hline 3.09 \end{array}$

5. $\begin{array}{r} \overset{5}{\cancel{6}}.{}^{1}0 \\ -0.3 \\ \hline 5.7 \end{array}$

6. $\begin{array}{r} \overset{2}{\cancel{3}}.{}^{1}7 \\ -2.8 \\ \hline 0.9 \end{array}$

7. $\begin{array}{r} {}^{0}\cancel{1}\cancel{1}.\overset{11}{\cancel{2}}{}^{1}0 \\ -8.23 \\ \hline 2.97 \end{array}$

8. $\begin{array}{r} {}^{0}\cancel{1}.\overset{9}{\cancel{0}}{}^{1}0 \\ -0.77 \\ \hline 0.23 \end{array}$

9. $\begin{array}{r} \overset{8}{\cancel{9}}.{}^{1}2 \\ -0.5 \\ \hline 8.7 \end{array}$

10. $\begin{array}{r} 8.1 \\ -7.0 \\ \hline 1.1 \end{array}$

11. $\begin{array}{r} 0.706 \\ -0.300 \\ \hline 0.406 \end{array}$

12. $\begin{array}{r} 0.2\overset{8}{\cancel{9}}{}^{1}8 \\ -0.009 \\ \hline 0.289 \end{array}$

13. $\begin{array}{r} {}^{1}\cancel{2}\overset{4}{\cancel{2}}5.\overset{9}{\cancel{0}}{}^{1}0 \\ -\$94.76 \\ \hline \$130.24 \end{array}$

14. $\begin{array}{r} \$\overset{9}{\cancel{1}}0.\overset{9}{\cancel{0}}{}^{1}0 \\ -\$7.45 \\ \hline \$2.55 \end{array}$

15. $\begin{array}{r} 6.7 \\ -4.3 \\ \hline 2.4 \end{array}$ loaves

16. $\begin{array}{r} \$\overset{1}{\cancel{2}}\overset{9}{\cancel{0}}.\overset{9}{\cancel{0}}{}^{1}0 \\ -\$16.25 \\ \hline \$3.75 \end{array}$

17. $\begin{array}{r} \overset{2}{\cancel{3}}\overset{10}{\cancel{1}}.{}^{1}0 \\ -7.5 \\ \hline 23.5 \end{array}$ chapters

18. $\begin{array}{r} \overset{3}{\cancel{4}}{}^{1}5.6 \\ -6.5 \\ \hline 39.1 \end{array}$ ft

Systematic Review 5D

1. $\begin{array}{r} \overset{3}{\cancel{4}}.{}^{1}2 \\ -3.9 \\ \hline 0.3 \end{array}$

2. $\begin{array}{r} 8.\overset{5}{\cancel{6}}{}^{1}0 \\ -0.04 \\ \hline 8.56 \end{array}$

3. $\begin{array}{r} 0.007 \\ -0.002 \\ \hline 0.005 \end{array}$

4. $\begin{array}{r} {}^{1}0.99 \\ +0.02 \\ \hline 1.01 \end{array}$

5. $\begin{array}{r} {}^{1}5.7 \\ +2.3 \\ \hline 8.0 \end{array}$

6. $\begin{array}{r} 6.025 \\ +0.800 \\ \hline 6.825 \end{array}$

7. $\frac{1}{6}+\frac{4}{6}=\frac{5}{6}$

8. $\frac{2}{9}+\frac{1}{10}=\frac{20}{90}+\frac{9}{90}=\frac{29}{90}$

9. $\frac{2}{5}+\frac{3}{8}=\frac{16}{40}+\frac{15}{40}=\frac{31}{40}$

10. $1\times1{,}000+3\times100+4\times\frac{1}{10}$

11. $1\times1+9\times\frac{1}{10}+7\times\frac{1}{100}+8\times\frac{1}{1{,}000}$

12. done

13. $\frac{2}{3}-\frac{1}{4}=\frac{8}{12}-\frac{3}{12}=\frac{5}{12}$

14. $\frac{5}{6}-\frac{3}{7}=\frac{35}{42}-\frac{18}{42}=\frac{17}{42}$

15. $\begin{array}{r} {}^9\!\!\cancel{10}.{}^1\!25 \\ -\;9.75 \\ \hline 0.50\text{mi} \end{array}$

16. $\begin{array}{r} \$15.00 \\ +\$21.75 \\ \hline \$36.75 \end{array}\qquad \begin{array}{r} \$3{}^5\!\cancel{6}.{}^{16}\!\cancel{7}\,{}^1\!5 \\ -\$31.99 \\ \hline \$\;\;4.76 \end{array}$

17. $\frac{2}{3}-\frac{1}{6}=\frac{12}{18}-\frac{3}{18}=\frac{9}{18}=\frac{1}{2}$ of a pie

18. $\frac{1}{2}=\frac{2}{4}$ They got the same amount.

Systematic Review 5E

1. $\begin{array}{r} {}^7\!\!\cancel{8}.{}^1\!6 \\ -2.70 \\ \hline 5.46 \end{array}$

2. $\begin{array}{r} {}^8\!\!\cancel{9}.{}^1\!04 \\ -2.90 \\ \hline 6.14 \end{array}$

3. $\begin{array}{r} 3.6\,{}^3\!\cancel{4}\,{}^1\!3 \\ -2.008 \\ \hline 1.635 \end{array}$

4. $\begin{array}{r} {}^1\!6.9 \\ +1.2 \\ \hline 8.1 \end{array}$

5. $\begin{array}{r} 4.007 \\ +0.932 \\ \hline 4.939 \end{array}$

6. $\begin{array}{r} {}^1\!31.600 \\ +\;\;0.456 \\ \hline 32.056 \end{array}$

7. $\frac{1}{8}+\frac{7}{9}=\frac{9}{72}+\frac{56}{72}=\frac{65}{72}$

8. $\frac{4}{5}+\frac{2}{11}=\frac{44}{55}+\frac{10}{55}=\frac{54}{55}$

9. $\frac{1}{3}+\frac{1}{4}=\frac{4}{12}+\frac{3}{12}=\frac{7}{12}$

10. $2\cdot 10^1+8\cdot 10^0+7\cdot\frac{1}{10^1}+8\cdot\frac{1}{10^2}$

11. $2\cdot 10^4$

12. $\frac{8}{9}-\frac{5}{9}=\frac{3}{9}=\frac{1}{3}$

13. $\frac{4}{5}-\frac{1}{2}=\frac{8}{10}-\frac{5}{10}=\frac{3}{10}$

14. $\frac{7}{10}-\frac{1}{6}=\frac{42}{60}-\frac{10}{60}=\frac{32}{60}=\frac{8}{15}$

15. $\begin{array}{r} 7.5\,{}^4\!\cancel{0} \\ -5.25 \\ \hline 2.25\text{ cups} \end{array}$

16. $\begin{array}{r} {}^1\!0.90 \\ 1.25 \\ 0.30 \\ +0.50 \\ \hline 2.95\text{ miles} \end{array}$

17. $\begin{array}{r} 4.5 \\ +13.2 \\ \hline 17.7\text{ inches cut off} \end{array}$

    $\begin{array}{r} {}^1\!\cancel{2}\,{}^{13}\!\cancel{4}.{}^1\!0 \\ -17.7 \\ \hline 6.3\text{ inches left} \end{array}$

18. $\frac{10}{100}=\frac{1}{10}$

19. $100\cent\div 10=10\cent;$
    $10\cent\times 5=50\cent$

20. $\frac{2}{5} + \frac{1}{2} = \frac{4}{10} + \frac{5}{10} = \frac{9}{10}$ are gone

$1 - \frac{9}{10} = \frac{10}{10} - \frac{9}{10} = \frac{1}{10}$ on the job

Systematic Review 5F

1. $\overset{7}{\cancel{8}}.\overset{1}{3}0$
 -1.9
 $\overline{6.4}$

2. $\overset{5}{\cancel{6}}.\overset{1}{1}8$
 -3.21
 $\overline{2.97}$

3. $7.\overset{0}{\cancel{1}}\overset{11}{\cancel{2}}\overset{1}{3}$
 -4.045
 $\overline{3.078}$

4. $\overset{1}{}6.1$
 $+0.9$
 $\overline{7.0}$

5. $\overset{1}{}3.930$
 $+0.605$
 $\overline{4.535}$

6. 28.700
 $+0.008$
 $\overline{28.708}$

7. $\frac{2}{4} + \frac{2}{5} = \frac{10}{20} + \frac{8}{20} = \frac{18}{20} = \frac{9}{10}$

8. $\frac{2}{3} + \frac{1}{4} = \frac{8}{12} + \frac{3}{12} = \frac{11}{12}$

9. $\frac{1}{6} + \frac{1}{5} = \frac{5}{30} + \frac{6}{30} = \frac{11}{30}$

10. 150,000.04

11. 6,800.22

12. $\frac{5}{8} - \frac{1}{3} = \frac{15}{24} - \frac{8}{24} = \frac{7}{24}$

13. $\frac{9}{10} - \frac{2}{5} = \frac{45}{50} - \frac{20}{50} = \frac{25}{50} = \frac{1}{2}$

14. $\frac{1}{4} - \frac{1}{7} = \frac{7}{28} - \frac{4}{28} = \frac{3}{28}$

15. $\overset{5}{\cancel{6}}.\overset{3}{\cancel{3}}\overset{1}{4}5$
 -4.738
 $\overline{1.607}$ ounces

16. $\overset{11}{}\$25.56$
 $+\$6.78$
 $\overline{\$32.34}$

17. $\$\overset{2}{\cancel{3}}\overset{2}{2}.\overset{2}{\cancel{3}}\overset{1}{4}$
 $-\$16.16$
 $\overline{\$16.18}$

18. $100¢ \div 100 = 1¢$;
 $1¢ \times 17 = 17¢$

19. $\frac{1}{3} + \frac{1}{4} = \frac{4}{12} + \frac{3}{12} = \frac{7}{12}$ of the candy

20. $36 \div 12 = 3$;
 $3 \times 7 = 21$ pieces given away
 $36 - 21 = 15$ pieces left

Lesson Practice 6A

1. b: meter
2. c: liter
3. a: gram
4. f: 10
5. d: 100
6. e: 1,000
7. $\frac{1 \text{ kilogram(kg)}}{1,000 \text{ grams (g)}}$; $\frac{1 \text{ hectogram(hg)}}{100 \text{ grams (g)}}$;
 $\frac{1 \text{ dekagram(dag)}}{10 \text{ grams (g)}}$; $\frac{1 \text{ gram(g)}}{1 \text{ gram(g)}}$
8. $\frac{1 \text{ kiloliter(kl)}}{1,000 \text{ liters (L)}}$; $\frac{1 \text{ hectoliter(hl)}}{100 \text{ liters (L)}}$;
 $\frac{1 \text{ dekaliter(dal)}}{10 \text{ liters (L)}}$; $\frac{1 \text{ liter(L)}}{1 \text{ liter(L)}}$
9. $\frac{1 \text{ kilometer(km)}}{1,000 \text{ meters (m)}}$; $\frac{1 \text{ hectometer(hm)}}{100 \text{ meters (m)}}$;
 $\frac{1 \text{ dekameter(dam)}}{10 \text{ meters (m)}}$; $\frac{1 \text{ meter(m)}}{1 \text{ meter(m)}}$
10. 1,000 meters
11. 100 liters
12. 10 grams

Lesson Practice 6B

1. meter
2. liter
3. gram
4. 1,000
5. 10
6. 100
7. $\dfrac{1 \text{ kilogram(kg)}}{1{,}000 \text{ grams (g)}}$; $\dfrac{1 \text{ hectogram(hg)}}{100 \text{ grams (g)}}$; $\dfrac{1 \text{ dekagram(dag)}}{10 \text{ grams (g)}}$; $\dfrac{1 \text{ gram(g)}}{1 \text{ gram(g)}}$
8. $\dfrac{1 \text{ kiloliter(kl)}}{1{,}000 \text{ liters (L)}}$; $\dfrac{1 \text{ hectoliter(hl)}}{100 \text{ liters (L)}}$; $\dfrac{1 \text{ dekaliter(dal)}}{10 \text{ liters (L)}}$; $\dfrac{1 \text{ liter(L)}}{1 \text{ liter(L)}}$
9. $\dfrac{1 \text{ kilometer(km)}}{1{,}000 \text{ meters (m)}}$; $\dfrac{1 \text{ hectometer(hm)}}{100 \text{ meters (m)}}$; $\dfrac{1 \text{ dekameter(dam)}}{10 \text{ meters (m)}}$; $\dfrac{1 \text{ meter(m)}}{1 \text{ meter(m)}}$
10. 100 grams
11. 1,000 liters
12. 10 meters

Lesson Practice 6C

1. c: meter
2. a: liter
3. b: gram
4. hecto
5. deka
6. kilo
7. $\dfrac{1 \text{ kilogram(kg)}}{1{,}000 \text{ grams (g)}}$; $\dfrac{1 \text{ hectogram(hg)}}{100 \text{ grams (g)}}$; $\dfrac{1 \text{ dekagram(dag)}}{10 \text{ grams (g)}}$; $\dfrac{1 \text{ gram(g)}}{1 \text{ gram(g)}}$
8. $\dfrac{1 \text{ kiloliter(kl)}}{1{,}000 \text{ liters (L)}}$; $\dfrac{1 \text{ hectoliter(hl)}}{100 \text{ liters (L)}}$; $\dfrac{1 \text{ dekaliter(dal)}}{10 \text{ liters (L)}}$; $\dfrac{1 \text{ liter(L)}}{1 \text{ liter(L)}}$
9. $\dfrac{1 \text{ kilometer(km)}}{1{,}000 \text{ meters (m)}}$; $\dfrac{1 \text{ hectometer(hm)}}{100 \text{ meters (m)}}$; $\dfrac{1 \text{ dekameter(dam)}}{10 \text{ meters (m)}}$; $\dfrac{1 \text{ meter(m)}}{1 \text{ meter(m)}}$
10. 10 liters
11. 100 meters
12. 1,000 grams

Systematic Review 6D

1. $\dfrac{1 \text{ kilogram(kg)}}{1{,}000 \text{ grams (g)}}$; $\dfrac{1 \text{ hectogram(hg)}}{100 \text{ grams (g)}}$; $\dfrac{1 \text{ dekagram(dag)}}{10 \text{ grams (g)}}$; $\dfrac{1 \text{ gram(g)}}{1 \text{ gram(g)}}$
2. $\dfrac{1 \text{ kiloliter(kl)}}{1{,}000 \text{ liters (L)}}$; $\dfrac{1 \text{ hectoliter(hl)}}{100 \text{ liters (L)}}$; $\dfrac{1 \text{ dekaliter(dal)}}{10 \text{ liters (L)}}$; $\dfrac{1 \text{ liter(L)}}{1 \text{ liter(L)}}$
3. $\dfrac{1 \text{ kilometer(km)}}{1{,}000 \text{ meters (m)}}$; $\dfrac{1 \text{ hectometer(hm)}}{100 \text{ meters (m)}}$; $\dfrac{1 \text{ dekameter(dam)}}{10 \text{ meters (m)}}$; $\dfrac{1 \text{ meter(m)}}{1 \text{ meter(m)}}$
4. $\begin{array}{r} \overset{1}{}1.8 \\ +4.5 \\ \hline 6.3 \end{array}$
5. $\begin{array}{r} 6.4 \\ -5.3 \\ \hline 1.1 \end{array}$
6. $\begin{array}{r} \overset{1}{}7.23 \\ +\,3.54 \\ \hline 10.77 \end{array}$
7. $\begin{array}{r} 8.15 \\ -0.13 \\ \hline 8.02 \end{array}$
8. $\dfrac{1}{6} + \dfrac{2}{3} = \dfrac{3}{18} + \dfrac{12}{18} = \dfrac{15}{18} = \dfrac{5}{6}$
9. $\dfrac{2}{5} + \dfrac{1}{2} = \dfrac{4}{10} + \dfrac{5}{10} = \dfrac{9}{10}$
10. $\dfrac{2}{3} - \dfrac{1}{4} = \dfrac{8}{12} - \dfrac{3}{12} = \dfrac{5}{12}$
11. $\dfrac{3}{5} - \dfrac{1}{3} = \dfrac{9}{15} - \dfrac{5}{15} = \dfrac{4}{15}$
12. $1\dfrac{5}{8} = \dfrac{13}{8}$
13. $4\dfrac{1}{2} = \dfrac{9}{2}$
14. $2\dfrac{3}{7} = \dfrac{17}{7}$
15. $5\dfrac{1}{3} = \dfrac{16}{3}$
16. $\begin{array}{r} \$5.95 \\ -\$1.50 \\ \hline \$4.45 \end{array}$
17. 1 km = 1,000 m

SYSTEMATIC REVIEW 6D - SYSTEMATIC REVIEW 6F

18. $\frac{7}{8} - \frac{1}{4} = \frac{28}{32} - \frac{8}{32} =$
    $\frac{20}{32} = \frac{5}{8}$ of a pizza

Systematic Review 6E

1. $\frac{1 \text{ kilogram(kg)}}{1,000 \text{ grams (g)}}$; $\frac{1 \text{ hectogram(hg)}}{100 \text{ grams (g)}}$;
   $\frac{1 \text{ dekagram(dag)}}{10 \text{ grams (g)}}$; $\frac{1 \text{ gram(g)}}{1 \text{ gram(g)}}$

2. $\frac{1 \text{ kiloliter(kl)}}{1,000 \text{ liters (L)}}$; $\frac{1 \text{ hectoliter(hl)}}{100 \text{ liters (L)}}$;
   $\frac{1 \text{ dekaliter(dal)}}{10 \text{ liters (L)}}$; $\frac{1 \text{ liter(L)}}{1 \text{ liter(L)}}$

3. $\frac{1 \text{ kilometer(km)}}{1,000 \text{ meters (m)}}$; $\frac{1 \text{ hectometer(hm)}}{100 \text{ meters (m)}}$;
   $\frac{1 \text{ dekameter(dam)}}{10 \text{ meters (m)}}$; $\frac{1 \text{ meter(m)}}{1 \text{ meter(m)}}$

4. 9.4
   +0.5

   9.9

5. $\overset{6}{\cancel{7}}.^10$
    -3.1

    3.9

6. $\overset{1\,1}{}$
    68.910
    + 2.306

    71.216

7. $0.4\overset{4}{\cancel{5}}\,^13$
    $-0.1\,2\,7$

    0.3 2 6

8. $\frac{2}{4} + \frac{1}{3} = \frac{6}{12} + \frac{4}{12} = \frac{10}{12} = \frac{5}{6}$

9. $\frac{2}{6} + \frac{1}{4} = \frac{8}{24} + \frac{6}{24} = \frac{14}{24} = \frac{7}{12}$

10. $\frac{3}{4} - \frac{1}{5} = \frac{15}{20} - \frac{4}{20} = \frac{11}{20}$

11. $\frac{4}{5} - \frac{1}{2} = \frac{8}{10} - \frac{5}{10} = \frac{3}{10}$

12. $2\frac{1}{3} = \frac{7}{3}$

13. $1\frac{1}{5} = \frac{6}{5}$

14. $6\frac{1}{2} = \frac{13}{2}$

15. $3\frac{4}{5} = \frac{19}{5}$

16. $5\frac{3}{4}$ dollars $= \frac{23}{4}$ dollars, or 23 quarters

17. $1 \times 10^2 + 7 \times 10^1 + 6 \times 10^0 + 4 \times \frac{1}{10^1}$

18. grams

19.   11.50
    + 12.25

    23.75 minutes spent by Rachel

    $\overset{2}{\cancel{3}}\overset{9}{\cancel{0}}.\overset{14}{\cancel{5}}\,^10$
    $-2\,3.\,7\,5$
    ---
    6. 7 5 more minutes Sierra spent

20. 8.875
    −6.625

    2.250 in

Systematic Review 6F

1. $\frac{1 \text{ kilogram(kg)}}{1,000 \text{ grams (g)}}$; $\frac{1 \text{ hectogram(hg)}}{100 \text{ grams (g)}}$;
   $\frac{1 \text{ dekagram(dag)}}{10 \text{ grams (g)}}$; $\frac{1 \text{ gram(g)}}{1 \text{ gram(g)}}$

2. $\frac{1 \text{ kiloliter(kl)}}{1,000 \text{ liters (L)}}$; $\frac{1 \text{ hectoliter(hl)}}{100 \text{ liters (L)}}$;
   $\frac{\text{dekaliter(dal)}}{10 \text{ liters (L)}}$; $\frac{1 \text{ liter(L)}}{1 \text{ liter(L)}}$

3. $\frac{1 \text{ kilometer(km)}}{1,000 \text{ meters (m)}}$; $\frac{1 \text{ hectometer(hm)}}{100 \text{ meters (m)}}$;
   $\frac{1 \text{ dekameter(dam)}}{10 \text{ meters (m)}}$; $\frac{1 \text{ meter(m)}}{1 \text{ meter(m)}}$

4.   6.11
    +0.05

    6.16

5.   4.8
    −1.0

    3.8

6. $\overset{1}{}$
    3.491
    +4.276

    7.767

7. $\overset{1}{\underset{}{2}}.\overset{9}{\underset{}{0}}\overset{9}{\underset{}{0}}7$
   -0.389
   _____
   1.618

8. $\frac{2}{6} + \frac{1}{5} = \frac{10}{30} + \frac{6}{30} = \frac{16}{30} = \frac{8}{15}$

9. $\frac{1}{2} + \frac{3}{9} = \frac{9}{18} + \frac{6}{18} = \frac{15}{18} = \frac{5}{6}$

10. $\frac{4}{7} - \frac{1}{4} = \frac{16}{28} - \frac{7}{28} = \frac{9}{28}$

11. $\frac{5}{6} - \frac{1}{8} = \frac{40}{48} - \frac{6}{48} = \frac{34}{48} = \frac{17}{24}$

12. $1\frac{1}{8} = \frac{9}{8}$

13. $3\frac{2}{5} = \frac{17}{5}$

14. $5\frac{1}{4} = \frac{21}{4}$

15. $7\frac{3}{10} = \frac{73}{10}$

16. $3\frac{5}{6} = \frac{23}{6}$ of a pie; 23 people

17. $5^3 = 5 \times 5 \times 5 = 125$

18. $\overset{1}{3}.6$
    $+0.8$
    _____
    4.4 rolls needed

    $\overset{4}{5}.^{1}0$
    -4.4
    _____
    0.6 roll left over

19. liters

20. $16 \div 4 = 4$;
    $4 \times 3 = 12$ children wanted to play
    $16 - 12 = 4$ children didn't want to play

Lesson Practice 7A

1. yard
2. quart
3. inch
4. f: $\frac{1}{10}$
5. e: $\frac{1}{100}$
6. d: $\frac{1}{1,000}$

7. $\frac{1\ gram(g)}{1\ gram(g)}$; $\frac{1\ decigram(dg)}{\frac{1}{10}\ gram(g)}$;
   $\frac{1\ centigram(cg)}{\frac{1}{100}\ gram(g)}$; $\frac{1\ milligram(mg)}{\frac{1}{1,000}\ gram(g)}$

8. $\frac{1\ liter(L)}{1\ liter(L)}$; $\frac{1\ deciliter(dl)}{\frac{1}{10}\ liter(L)}$;
   $\frac{1\ centiliter(cl)}{\frac{1}{100}\ liter(L)}$; $\frac{1\ milliliter(ml)}{\frac{1}{1,000}\ liter(L)}$

9. $\frac{1\ meter(m)}{1\ meter(m)}$; $\frac{1\ decimeter(dm)}{\frac{1}{10}\ meter(m)}$;
   $\frac{1\ centimeter(cm)}{\frac{1}{100}\ meter(m)}$; $\frac{1\ millimeter(mm)}{\frac{1}{1,000}\ meter(m)}$

10. $100\ cm = 1\ m$
11. 1 deciliter
12. $28\ g \times 16\ g = 448$ grams

Lesson Practice 7B

1. gram
2. kilogram
3. kilometers
4. $\frac{1}{100}$
5. $\frac{1}{10}$
6. $\frac{1}{1,000}$

7. $\frac{1\ gram(g)}{1\ gram(g)}$; $\frac{1\ decigram(dg)}{\frac{1}{10}\ gram(g)}$;
   $\frac{1\ centigram(cg)}{\frac{1}{100}\ gram(g)}$; $\frac{1\ milligram(mg)}{\frac{1}{1,000}\ gram(g)}$

8. $\frac{1\ liter(L)}{1\ liter(L)}$; $\frac{1\ deciliter(dl)}{\frac{1}{10}\ liter(L)}$;
   $\frac{1\ centiliter(cl)}{\frac{1}{100}\ liter(L)}$; $\frac{1\ milliliter(ml)}{\frac{1}{1,000}\ liter(L)}$

LESSON PRACTICE 7B - SYSTEMATIC REVIEW 7D

9. $\frac{1\ meter(m)}{1\ meter(m)}$; $\frac{1\ decimeter(dm)}{\frac{1}{10}\ meter(m)}$;

   $\frac{1\ centimeter(cm)}{\frac{1}{100}\ meter(m)}$; $\frac{1\ millimeter(mm)}{\frac{1}{1,000}\ meter(m)}$

10. 1 milliliter
11. 1 centimeter
12. grams

Lesson Practice 7C
1. meters
2. 2 miles
3. milliliters
4. grams
5. 16 ounces
6. 36 inches
7. $\frac{1\ gram(g)}{1\ gram(g)}$; $\frac{1\ decigram(dg)}{\frac{1}{10}\ gram(g)}$;

   $\frac{1\ centigram(cg)}{\frac{1}{100}\ gram(g)}$; $\frac{1\ milligram(mg)}{\frac{1}{1,000}\ gram(g)}$

8. $\frac{1\ liter(L)}{1\ liter(L)}$; $\frac{1\ deciliter(dl)}{\frac{1}{10}\ liter(L)}$;

   $\frac{1\ centiliter(cl)}{\frac{1}{100}\ liter(L)}$; $\frac{1\ milliliter(ml)}{\frac{1}{1,000}\ liter(L)}$

9. $\frac{1\ meter(m)}{1\ meter(m)}$; $\frac{1\ decimeter(dm)}{\frac{1}{10}\ meter(m)}$;

   $\frac{1\ centimeter(cm)}{\frac{1}{100}\ meter(m)}$; $\frac{1\ millimeter(mm)}{\frac{1}{1,000}\ meter(m)}$

10. 1 millimeter
11. 1 liter
12. 50 grams

Systematic Review 7D
1. $\frac{1\ gram(g)}{1\ gram(g)}$; $\frac{1\ decigram(dg)}{\frac{1}{10}\ gram(g)}$;

   $\frac{1\ centigram(cg)}{\frac{1}{100}\ gram(g)}$; $\frac{1\ milligram(mg)}{\frac{1}{1,000}\ gram(g)}$

2. $\frac{1\ liter(L)}{1\ liter(L)}$; $\frac{1\ deciliter(dl)}{\frac{1}{10}\ liter(L)}$;

   $\frac{1\ centiliter(cl)}{\frac{1}{100}\ liter(L)}$; $\frac{1\ milliliter(ml)}{\frac{1}{1,000}\ liter(L)}$

3. $\frac{1\ meter(m)}{1\ meter(m)}$; $\frac{1\ decimeter(dm)}{\frac{1}{10}\ meter(m)}$;

   $\frac{1\ centimeter(cm)}{\frac{1}{100}\ meter(m)}$; $\frac{1\ millimeter(mm)}{\frac{1}{1,000}\ meter(m)}$

4. c: 1,000
5. b: 100
6. a: 10
7. $16 \div 2 = 8$
   $8 \times 1 = 8$
8. $40 \div 5 = 8$
   $8 \times 3 = 24$
9. $63 \div 9 = 7$
   $7 \times 2 = 14$
10. $48 \div 4 = 12$
    $12 \times 1 = 12$
11. $\frac{25}{8} = 3\frac{1}{8}$
12. $\frac{3}{2} = 1\frac{1}{2}$
13. $\frac{11}{9} = 1\frac{2}{9}$
14. $\frac{10}{3} = 3\frac{1}{3}$
15. liter
16. meter
17. $\overset{1}{7}.2$
    $+ 6.3$
    $\overline{13.5}$ tons brought by the first two trucks

    $17.8\ 0$
    $-13.5\ 0$
    $\overline{4.3\ 0}$ tons needed
18. kilometers

Systematic Review 7E

1. $\dfrac{1\ \text{gram(g)}}{1\ \text{gram(g)}}$; $\dfrac{1\ \text{decigram(dg)}}{\frac{1}{10}\ \text{gram(g)}}$;

   $\dfrac{1\ \text{centigram(cg)}}{\frac{1}{100}\ \text{gram(g)}}$; $\dfrac{1\ \text{milligram(mg)}}{\frac{1}{1{,}000}\ \text{gram(g)}}$

2. $\dfrac{1\ \text{liter(L)}}{1\ \text{liter(L)}}$; $\dfrac{1\ \text{deciliter(dl)}}{\frac{1}{10}\ \text{liter(L)}}$;

   $\dfrac{1\ \text{centiliter(cl)}}{\frac{1}{100}\ \text{liter(L)}}$; $\dfrac{1\ \text{milliliter(ml)}}{\frac{1}{1{,}000}\ \text{liter(L)}}$

3. $\dfrac{1\ \text{meter(m)}}{1\ \text{meter(m)}}$; $\dfrac{1\ \text{decimeter(dm)}}{\frac{1}{10}\ \text{meter(m)}}$;

   $\dfrac{1\ \text{centimeter(cm)}}{\frac{1}{100}\ \text{meter(m)}}$; $\dfrac{1\ \text{millimeter(mm)}}{\frac{1}{1{,}000}\ \text{meter(m)}}$

4. deka
5. hecto
6. kilo
7. $30 \div 3 = 10$
   $10 \times 1 = 10$
8. $12 \div 6 = 2$
   $2 \times 5 = 10$
9. $49 \div 7 = 7$
   $7 \times 1 = 7$
10. $64 \div 8 = 8$
    $8 \times 5 = 40$
11. $\dfrac{37}{9} = 4\dfrac{1}{9}$
12. $\dfrac{17}{4} = 4\dfrac{1}{4}$
13. $\dfrac{47}{10} = 4\dfrac{7}{10}$
14. $\dfrac{53}{5} = 10\dfrac{3}{5}$
15. centimeters
16. 1,000
17. 1 mile
18. kilogram

19.
```
  2.50
 +3.25
 -----
  5.75 gallons used
```

```
  7.50
 -5.75
 -----
  1.75 gallons left
```

20. $\dfrac{1}{2} - \dfrac{1}{6} = \dfrac{6}{12} - \dfrac{2}{12} = \dfrac{4}{12} = \dfrac{1}{3}$ of the job

Systematic Review 7F

1. $\dfrac{1\ \text{gram(g)}}{1\ \text{gram(g)}}$; $\dfrac{1\ \text{decigram(dg)}}{\frac{1}{10}\ \text{gram(g)}}$;

   $\dfrac{1\ \text{centigram(cg)}}{\frac{1}{100}\ \text{gram(g)}}$; $\dfrac{1\ \text{milligram(mg)}}{\frac{1}{1{,}000}\ \text{gram(g)}}$

2. $\dfrac{1\ \text{liter(L)}}{1\ \text{liter(L)}}$; $\dfrac{1\ \text{deciliter(dl)}}{\frac{1}{10}\ \text{liter(L)}}$;

   $\dfrac{1\ \text{centiliter(cl)}}{\frac{1}{100}\ \text{liter(L)}}$; $\dfrac{1\ \text{milliliter(ml)}}{\frac{1}{1{,}000}\ \text{liter(L)}}$

3. $\dfrac{1\ \text{meter(m)}}{1\ \text{meter(m)}}$; $\dfrac{1\ \text{decimeter(dm)}}{\frac{1}{10}\ \text{meter(m)}}$;

   $\dfrac{1\ \text{centimeter(cm)}}{\frac{1}{100}\ \text{meter(m)}}$; $\dfrac{1\ \text{millimeter(mm)}}{\frac{1}{1{,}000}\ \text{meter(m)}}$

4. dekagram
5. hectoliter
6. kilometer
7. $72 \div 8 = 9$
   $9 \times 3 = 27$
8. $50 \div 2 = 25$
   $25 \times 1 = 25$
9. $24 \div 3 = 8$
   $8 \times 2 = 16$
10. $44 \div 11 = 4$
    $4 \times 4 = 16$
11. $\dfrac{28}{6} = 4\dfrac{4}{6} = 4\dfrac{2}{3}$
12. $\dfrac{11}{7} = 1\dfrac{4}{7}$
13. $\dfrac{16}{5} = 3\dfrac{1}{5}$

14. $\frac{31}{2} = 15\frac{1}{2}$
15. the meter stick
16. 10 pounds
17. liters
18. $\$\ 6.15$
    $\$\ 8.45$
    $+\$10.25$
    $\$24.85$
19. meters
20. $780 \div 10 = 78$
    $78 \times 3 = 234$ people

Lesson Practice 8A

1. done
2. $1\ \cancel{dam} \times \frac{10\ \cancel{m}}{1\ \cancel{dam}} \times \frac{100\ cm}{1\ \cancel{m}} = 1,000\ cm;$
   three steps: $1(10)(10)(10) = 1,000\ cm$
3. 10 ml
4. 1,000 dl
5. 1,000 mg
6. 100 mg
7. done
8. 3,000,000 mm
9. 3,500 ml
10. 1,300 g
11. 600,000 cl
12. 1,000 dg
13. 60,000 cg
14. 100,000 dm
15. 90 pieces

Lesson Practice 8B

1. 10,000 mm
2. 1,000,000 ml
3. 10 cl
4. 10 hg
5. 1,000 dm
6. 100 mm
7. 80,000 cm
8. 2,000 cl
9. 6,000,000 mg
10. 3,200 dg
11. 90 hl
12. 2,200,000 cm
13. 3,000 ml
14. 4,500 dg
15. 1,200 cm

Lesson Practice 8C

1. 100,000 cm
2. 10 dam
3. 10,000 mg
4. 10,000 cg
5. 10,000 dl
6. 10 L
7. 1,000 dam
8. 150,000 cg
9. 700 cm
10. 60 mm
11. 5,000 ml
12. 24,000 g
13. 500,000 seconds
14. 80 pieces
15. 2 liters = 2,000 ml
    2,000 ml > 1,983 ml
    2 liters is more

Systematic Review 8D

1. 500 cg
2. 10,000 mm
3. 160,000 cm
4. c: 1,000
5. b: 100
6. a: 10
7. d: $\frac{1}{10}$
8. f: $\frac{1}{100}$
9. e: $\frac{1}{1,000}$
10. $2\frac{3}{8} = \frac{19}{8}$
11. $6\frac{1}{7} = \frac{43}{7}$

12. $\frac{17}{9} = 1\frac{8}{9}$

13. $\frac{8}{5} = 1\frac{3}{5}$

14. done

15. $3\frac{3}{8} + 1\frac{4}{5} = 3\frac{15}{40} + 1\frac{32}{40} = 4\frac{47}{40} =$
    $4 + \frac{40}{40} + \frac{7}{40} = 4 + 1 + \frac{7}{40} = 5\frac{7}{40}$

16. $2\frac{1}{10} + 3\frac{5}{8} = 2\frac{8}{80} + 3\frac{50}{80} = 5\frac{58}{80} = 5\frac{29}{40}$

17. 2,000 g

18. $1\frac{1}{2}\text{hr} + 1\frac{1}{4}\text{hr} = 1\frac{4}{8}\text{hr} + 1\frac{2}{8}\text{hr}$
    $= 2\frac{6}{8}\text{hr} = 2\frac{3}{4}\text{hr}$

19. $\begin{array}{r} \overset{31014}{41.50} \\ -\ 3\ 7.\ 7\ 5 \\ \hline 3.\ 7\ 5\text{ in} \end{array}$

20. $\frac{1}{3}$ of 900 = 300 in tanks

    $\frac{1}{6}$ of 900 = 150 in trucks

    300 + 150 = 450 riding
    900 − 450 = 450 walking

    You could also have added the fractions and used your answer to find the number of soldiers who were riding. There is often more than one way to solve a word problem.

Systematic Review 8E

1. 32,000 ml
2. 90,000 ml
3. 150 mm
4. b: 1 yard
5. a: 0.4 of an inch
6. c: 0.6 of a mile
7. e: 1 quart
8. f: $\frac{1}{500}$ of a pound
9. d: 2.2 pounds
10. $5\frac{1}{8} = \frac{41}{8}$
11. $4\frac{2}{3} = \frac{14}{3}$
12. $\frac{22}{7} = 3\frac{1}{7}$
13. $\frac{27}{5} = 5\frac{2}{5}$
14. $8\frac{2}{3} + 5\frac{1}{4} = 8\frac{8}{12} + 5\frac{3}{12} = 13\frac{11}{12}$
15. $18\frac{1}{10} + 3\frac{5}{8} = 18\frac{8}{80} + 3\frac{50}{80} =$
    $21\frac{58}{80} = 21\frac{29}{40}$
16. $11\frac{3}{5} + 4\frac{1}{6} = 11\frac{18}{30} + 4\frac{5}{30} = 15\frac{23}{30}$
17. 100 cm = 10 dm; same length
18. 1 km

Systematic Review 8F

1. 400,000 cg
2. 800 dag
3. 5,000,000 ml
4. b: length
5. c: volume
6. a: weight
7. d: length
8. f: weight
9. e: volume
10. $2\frac{3}{5} = \frac{13}{5}$
11. $1\frac{1}{7} = \frac{8}{7}$
12. $\frac{19}{8} = 2\frac{3}{8}$
13. $\frac{31}{6} = 5\frac{1}{6}$
14. $9\frac{1}{3} + 6\frac{1}{4} = 9\frac{4}{12} + 6\frac{3}{12} = 15\frac{7}{12}$
15. $4\frac{2}{3} + 1\frac{1}{5} = 4\frac{10}{15} + 1\frac{3}{15} = 5\frac{13}{15}$
16. $12\frac{2}{10} + 4\frac{5}{8} = 12\frac{16}{80} + 4\frac{50}{80} =$
    $16\frac{66}{80} = 16\frac{33}{40}$
17. 45,000 g
    45,000,000 mg
18. $5 \times 10 = 50$ bottles

19. $1\times10^1 + 2\times10^0 + 8\times\dfrac{1}{10^1} + 7\times\dfrac{1}{10^2} + 4\times\dfrac{1}{10^3}$

20.   1
   6.20
   4.45
+ 9.00
 19.65 ft

Lesson Practice 9A

1. done
2. done
3.   1 + 0.4   1.4
 × 1 + 0.2   ×1.2
   0.2 + 0.08   0.28
   1 + 0.4   1.4
   1 + 0.6 + 0.08   1.68

4.   1 + 0.1   1.1
 ×   0.6   ×0.6
   0.6 + 0.06   0.66

5.   1 + 0.2   1.2
 × 2 + 0.1   ×2.1
   0.1 + 0.02   0.12
   2 + 0.4   2.4
   2 + 0.5 + 0.02   2.52

6.   2 + 0.2   2.2
 ×   0.3   ×0.3
   0.6 + 0.06   0.66

7.   1 + 0.3   1.3
 × 1 + 0.1   ×1.1
   0.1 + 0.03   0.13
   1 + 0.3   1.3
   1 + 0.4 + 0.03   1.43

8.   2 + 0.3   2.3
 ×   0.2   ×0.2
   0.4 + 0.06   0.46

9.   3.1
 ×0.3
  0.93 yd

10.   2.1
 ×0.4
  0.84 gal

Lesson Practice 9B

1.   2 + 0.8   2.8
 × 1 + 0.0   ×1.0
   0.0 + 0.00   0.00
   2 + 0.8   2.8
   2 + 0.8 + 0.00   2.80

2.   2 + 0.3   2.3
 ×   0.1   ×0.1
   0.2 + 0.03   0.23

3.   2 + 0.1   2.1
 × 1 + 0.4   ×1.4
   0.8 + 0.04   0.84
   2 + 0.1   2.1
   2 + 0.9 + 0.04   2.94

4.   1 + 0.1   1.1
 ×   0.8   ×0.8
   0.8 + 0.08   0.88

5.   2 + 0.2   2.2
 × 1 + 0.2   ×1.2
   0.4 + 0.04   0.44
   2 + 0.2   2.2
   2 + 0.6 + 0.04   2.64

6.   1 + 0.3   1.3
 ×   0.3   ×0.3
   0.3 + 0.09   0.39

7.   1 + 0.3   1.3
 × 1 + 0.3   ×1.3
   0.3 + 0.09   0.39
   1 + 0.3   1.3
   1 + 0.6 + 0.09   1.69

8.   1 + 0.1   1.1
 ×   0.9   ×0.9
   0.9 + 0.09   0.99

9.   2.6
 ×1.1
  0.26
  2.6
  2.86

10.   1.1
 ×0.5
  0.55 mi

Lesson Practice 9C

1. $\begin{array}{r} 1+0.3 \\ \times 2+0.2 \\ \hline 0.2+0.06 \\ 2+0.6 \\ \hline 2+0.8+0.06 \end{array}$ $\begin{array}{r} 1.3 \\ \times 2.2 \\ \hline 0.26 \\ 2.6 \\ \hline 2.86 \end{array}$

2. $\begin{array}{r} 2+0.0 \\ \times 0.1 \\ \hline 0.2+0.00 \end{array}$ $\begin{array}{r} 2.0 \\ \times 0.1 \\ \hline 0.2 \end{array}$

3. $\begin{array}{r} 1+0.3 \\ \times 1+0.2 \\ \hline 0.2+0.06 \\ 1+0.3 \\ \hline 1+0.5+0.06 \end{array}$ $\begin{array}{r} 1.3 \\ \times 1.2 \\ \hline 0.26 \\ 1.3 \\ \hline 1.56 \end{array}$

4. $\begin{array}{r} 2+0.1 \\ \times 0.6 \\ \hline 1+0.2+0.06 \end{array}$ $\begin{array}{r} 2.1 \\ \times 0.6 \\ \hline 1.26 \end{array}$

5. $\begin{array}{r} 2+0.1 \\ \times 1+0.4 \\ \hline 0.8+0.04 \\ 2+0.1 \\ \hline 2+0.9+0.04 \end{array}$ $\begin{array}{r} 2.1 \\ \times 1.4 \\ \hline 0.84 \\ 2.1 \\ \hline 2.94 \end{array}$

6. $\begin{array}{r} 2+0.1 \\ \times 0.4 \\ \hline 0.8+0.04 \end{array}$ $\begin{array}{r} 2.1 \\ \times 0.4 \\ \hline 0.84 \end{array}$

7. $\begin{array}{r} 2+0.3 \\ \times 1+0.1 \\ \hline 0.2+0.03 \\ 2+0.3 \\ \hline 2+0.5+0.03 \end{array}$ $\begin{array}{r} 2.3 \\ \times 1.1 \\ \hline 0.23 \\ 2.3 \\ \hline 2.53 \end{array}$

8. $\begin{array}{r} 0.3 \\ \times 0.3 \\ \hline 0.09 \end{array}$ $\begin{array}{r} 0.3 \\ \times 0.3 \\ \hline 0.09 \end{array}$

9. $\begin{array}{r} 3.2 \text{ miles/hour} \\ \times 0.1 \text{ hour} \\ \hline 0.32 \text{ miles} \end{array}$

10. $\begin{array}{r} 0.4 \\ \times 0.2 \\ \hline 0.08 \text{ of the jelly beans} \end{array}$

Systematic Review 9D

1. $\begin{array}{r} 2.5 \\ \times 1.1 \\ \hline 0.25 \\ 2.5 \\ \hline 2.75 \end{array}$

2. $\begin{array}{r} 2.3 \\ \times 1.3 \\ \hline 0.69 \\ 2.3 \\ \hline 2.99 \end{array}$

3. $\begin{array}{r} 1.1 \\ \times 0.5 \\ \hline 0.55 \end{array}$

4. $\begin{array}{r} 0.2 \\ \times 0.2 \\ \hline 0.04 \end{array}$

5. 900 mg
6. 8,000 cl
7. 2,400 cg
8. cg
9. ml
10. km
11. dal
12. dl
13. kg
14. done
15. done
16. $5\frac{1}{4} - 1\frac{2}{3} = 5\frac{3}{12} - 1\frac{8}{12} = 4\frac{15}{12} - 1\frac{8}{12} = 3\frac{7}{12}$
17. $3\frac{1}{2}$ yd $- \frac{1}{3}$ yd $= 3\frac{3}{6}$ yd $- \frac{2}{6}$ yd $= 3\frac{1}{6}$ yards
18. $3\frac{3}{10} = \frac{33}{10}$

    $\frac{3}{10} \times \frac{33}{10} = \frac{99}{100}$ pies eaten

    $3\frac{3}{10} = 3\frac{30}{100} + \frac{100}{100} = \frac{130}{100}$

    $-\frac{99}{100} = -\frac{99}{100}$

    $2\frac{31}{100}$ OR 2.31 pies left over

Systematic Review 9E

1. 2.4
    ×0.2
    ―――
    0.48

2. 1.8
    ×1.1
    ―――
    0.18
    1.8
    ―――
    1.98

3. 1.7
    ×0.1
    ―――
    0.17

4. 0.1
    ×0.1
    ―――
    0.01

5. 280,000 cm
6. 3,600 dm
7. 500 dal
8. meter
9. quart
10. gram
11. $7\frac{2}{5} - 3\frac{1}{3} = 7\frac{6}{15} - 3\frac{5}{15} = 4\frac{1}{15}$
12. $9 - 2\frac{1}{2} = 8\frac{2}{2} - 2\frac{1}{2} = 6\frac{1}{2}$
13. $6\frac{1}{2} - 4\frac{1}{7} = 6\frac{7}{14} - 4\frac{2}{14} = 2\frac{5}{14}$
14. $\$7\frac{1}{2} - \$1\frac{1}{4} = \$7\frac{2}{4} - \$1\frac{1}{4}$
    $= 6\frac{1}{4}$ dollars or $6.25

    If one fraction can be made into an equivalent fraction with the same denominator, you don't have to use the Rule of Four. The final result will be the same.

15. 4.4 hours
    ×0.1 pounds/hour
    ―――――――
    0.44 pounds

16. ¹ ¹
    45.15
    34.10
    + 7.05
    ――――
    86.30 pounds

17. $6 \text{ hm} \times \frac{100 \text{ m}}{1 \text{ hm}} \times \frac{10 \text{ dm}}{1 \text{ m}} = 6,000 \text{ dm}$;
    she took 6,000 steps.

18. A meter is a little longer than a yard, so he ran faster in the 100-meter dash.

Systematic Review 9F

1. 2.3
    ×0.3
    ―――
    0.69

2. 2.1
    ×1.3
    ―――
    0.63
    2.1
    ―――
    2.73

3. 2.2
    ×0.4
    ―――
    0.88

4. 0.4
    ×0.2
    ―――
    0.08

5. 140 mm
6. 2,000 g
7. 110 m
8. pounds
9. ounce
10. kilometer
11. $10\frac{1}{3} - 4\frac{1}{4} = 10\frac{4}{12} - 4\frac{3}{12} = 6\frac{1}{12}$
12. $6 - 1\frac{2}{3} = 5\frac{3}{3} - 1\frac{2}{3} = 4\frac{1}{3}$
13. $13\frac{1}{5} - 8\frac{5}{6} = 13\frac{6}{30} - 8\frac{25}{30} =$
    $12\frac{36}{30} - 8\frac{25}{30} = 4\frac{11}{30}$
14. $1 \text{ L} \times \frac{1,000 \text{ ml}}{1 \text{ L}} = 1,000 \text{ ml}$;
    about 1,000 g
15. $1,000 \text{ g} \times \frac{1 \text{ kg}}{1,000 \text{ g}} = 1 \text{ kg}$
16. 1.1 pounds
    ×0.5
    ―――――
    0.55 pounds

17. $8 \text{ kg} \times \dfrac{1{,}000 \text{ g}}{1 \text{ kg}} = 8{,}000 \text{ g}$

18.
```
      6
    ⁷.¹0
   -2. 3
   ─────
    4. 7 minutes
```

Lesson Practice 10A

1. done
2. done
3.
```
    35.              35    0 places
   × 0.26          ×  26   2 places
   ──────          ──────
    2.10            2 10
    7.0             7 0
   ──────          ──────
    9.10            9.10   2 places
```
4.
```
    1.03            1 0 3   2 places
   ×0.76           ×   76   2 places
   ──────          ───────
   0.0618            6 18
   0.721             72 1
   ──────          ───────
   0.7828           0.7828  4 places
```
5.
```
    4.7             47    1 place
   ×0.6            ×  6   1 place
   ─────           ─────
    2.82            2.82  2 places
```
6.
```
    0.52            52    2 places
   ×0.14          ×  14   2 places
   ──────         ──────
   0.0208           208
   0.052             52
   ──────         ──────
   0.0728          0.0728  4 places
```
7.
```
    200.            200   0 places
   × 0.08          x  08   2 places
   ──────         ──────
   16.00           16.00   2 places
```
8.
```
    5.28            528    2 places
   ×0.12          ×  12    2 places
   ──────         ───────
   0.1056           1056
   0.528             528
   ──────         ───────
   0.6336          0.6336  4 places
```
9. $\$6.55/\text{hr} \times 0.4 \text{ hr} = \$2.62$
10. $\$0.96/\text{doz} \times 2.5 \text{ doz} = \$2.40$

(Note: The final zero is included for money.)

Lesson Practice 10B

1.
```
    3.4             34    1 place
   ×0.94          ×  94    2 places
   ──────         ──────
   0.136            136
   3.06             306
   ──────         ──────
   3.196           3.196   3 places
```
2.
```
    0.73            73    2 places
   ×0.48          ×  48    2 places
   ──────         ──────
   0.0584          0584
   0.292            292
   ──────         ──────
   0.3504          0.3504  4 places
```
3.
```
    33.             33    0 places
   × 0.42         ×  42    2 places
   ──────         ──────
    0.66             66
   13.2             13 2
   ──────         ──────
   13.86           13.86   2 places
```
4.
```
    5.19            5 19   2 places
   ×0.81          ×   81   2 places
   ──────         ───────
   0.0519           5 19
   4.152           4 152
   ──────         ───────
   4.2039          4.2039  4 places
```
5.
```
    9.6             96    1 place
   ×0.9           ×  9    1 place
   ─────          ─────
    8.64            8.64   2 places
```
6.
```
    0.64            64    2 places
   ×0.94          ×  94    2 places
   ──────         ──────
   0.0256           256
   0.576            576
   ──────         ──────
   0.6016          0.6016  4 places
```
7.
```
    116.            116   0 places
   × 0.02         ×  02    2 places
   ──────         ──────
    2.32            2.32   2 places
```
8.
```
    7.18            718    2 places
   ×0.05          x    5    2 places
   ──────         ───────
   0.3590          0.3590  4 places
```
9. $0.15 \text{ oz} \times 100 = 15 \text{ ounces}$
10. $0.33 \text{ g} \times 45 = 14.85 \text{ grams}$

Lesson Practice 10C

1.
```
   4.8           48   1 place
  ×0.71         ×  71  2 places
  ─────         ──────
  0.048          048
  3.36          3 36
  ─────         ──────
  3.408         3.408  3 places
```

2.
```
   0.62           62   2 places
  ×0.37          ×  37  2 places
  ──────         ──────
  0.0434          434
  0.186           186
  ──────         ──────
  0.2294         0.2294  4 places
```

3.
```
   69.            69   0 places
  ×  2.3         ×  23   1 place
  ──────         ──────
   20.7           20 7
  138            138
  ──────         ──────
  158.7          158.7   1 place
```

4.
```
   9.97           997   2 places
  ×0.11          ×  11   2 places
  ──────         ──────
  0.0997          997
  0.997           997
  ──────         ──────
  1.0967         1.0967   4 places
```

5.
```
   1.7            17   1 place
  ×0.4          ×   4   1 place
  ─────         ──────
  0.68          0.68   2 places
```

6.
```
   0.16           16   2 places
  ×0.54          ×  54   2 places
  ──────         ──────
  0.0064          64
  0.080           80
  ──────         ──────
  0.0864         0.0864   4 places
```

7.
```
   400.           4 00   0 places
  ×  0.11        ×  11   2 places
  ──────         ──────
   44.00          44.00   2 places
```

8.
```
   6.73           673   2 places
  ×0.46          ×  46   2 places
  ──────         ──────
  0.4038          4038
  2.692           2692
  ──────         ──────
  3.0958         3.0958   4 places
```

9. 4.25 bushels × 0.75 = 3.1875 bushel
10. 2.45 meters × 5 = 12.25 meters

Systematic Review 10D

1.
```
   2.6            26   1 place
  ×0.24          ×  24   2 places
  ──────         ──────
  0.104           104
  0.52             52
  ──────         ──────
  0.624          0.624   3 places
```

2.
```
   6.3            63   1 place
  × 5.7          ×  57   1 place
  ──────         ──────
   4.41           4 41
  31.5            31 5
  ──────         ──────
  35.91          35.91   2 places
```

3.
```
   3.52           352   2 places
  ×0.04          ×   04   2 places
  ──────         ──────
  0.1408         0.1408   4 places
```

4.
```
   0.67           67   2 places
  ×0.05          ×  05   2 places
  ──────         ──────
  0.0335         0.0335   4 places
```

5. 14,000 dg
6. 50 cm
7. $1.00 - 0.12 = 0.88$
8. $4.08 - 2.9 = 1.18$
9. $0.95 + 3.61 = 4.56$
10. $4\frac{3}{4} + 2\frac{1}{5} = 4\frac{15}{20} + 2\frac{4}{20} = 6\frac{19}{20}$
11. $12\frac{4}{7} - 6\frac{1}{7} = 6\frac{3}{7}$
12. $5\frac{1}{10} + 3\frac{2}{5} = 5\frac{5}{50} + 3\frac{20}{50} = 8\frac{25}{50} = 8\frac{1}{2}$
    or
    $5\frac{1}{10} + 3\frac{4}{10} = 8\frac{5}{10} = 8\frac{1}{2}$
13. $\frac{1}{\cancel{2}} \times \frac{\cancel{2}}{3} = \frac{1}{3}$
14. $\frac{3}{4} \times \frac{1}{\cancel{9}_3} = \frac{1}{12}$
15. $\frac{\cancel{6}^3}{8} \times \frac{1}{\cancel{2}} = \frac{3}{8}$
16. $\frac{\cancel{2}}{3} \times \frac{1}{\cancel{2}} = \frac{1}{3}$ of the package
    Writing "1" when canceling is optional.
17. 0.4 in/cm × 18 cm = 7.2 in
18. 55 kg × 2.2 lb/kg = 121 lb

Systematic Review 10E

1.
```
   11.9         1 19    1 place
 ×  5.3       ×   53    1 place
   3.57         3 57
  59.5         59 5
  63.07        63.07    2 places
```

2.
```
  0.05          05      2 places
× 0.4        ×   4      1 place
  0.02         0.020    3 places
```

3.
```
  1.07         107      2 places
× 0.72       ×  72      2 places
  0.0214       214
  0.749        749
  0.7704       0.7704   4 places
```

4.
```
  0.46          46      2 places
× 0.49       ×  49      2 places
  0.0414       414
  0.184        184
  0.2254       0.2254   4 places
```

5. 1,700 dg
6. 800,000 mm
7. $12.51 - 2.74 = 9.77$
8. $2.38 + 4.509 = 6.889$
9. $0.412 - 0.367 = 0.045$
10. $5\frac{3}{7} + 5\frac{1}{4} = 5\frac{12}{28} + 5\frac{7}{28} = 10\frac{19}{28}$
11. $2\frac{1}{8} - 1\frac{5}{9} = 2\frac{9}{72} - 1\frac{40}{72} = 1\frac{81}{72} - 1\frac{40}{72} = \frac{41}{72}$
12. $3\frac{1}{2} + 2\frac{1}{2} = 5\frac{2}{2} = 6$
13. $\frac{1}{2} \times \frac{3}{5} = \frac{3}{10}$
14. $\frac{3}{4} \times \frac{1}{7} = \frac{3}{28}$
15. $\frac{\cancel{4}^2}{5} \times \frac{7}{\cancel{10}_5} = \frac{14}{25}$
16. $\frac{3}{4} \div \frac{6}{1} = \frac{3}{4} \times \frac{1}{\cancel{6}_2} = \frac{1}{8}$
17. $1\frac{1}{6}\text{ yd} + 5\frac{3}{8}\text{ yd} = 1\frac{8}{48}\text{ yd} + 5\frac{18}{48}\text{ yd}$
    $= 6\frac{26}{48}\text{ yd} = 6\frac{13}{24}\text{ yd}$
18. 2.5 cm/in × 36 in = 90 cm
19. 1.06 qt/L × 8 L = 8.48 qt
20. 34,000 grams

Systematic Review 10F

1.
```
  9.35         9 35     2 places
×0.5         ×    5     1 place
  4.675        4.675    3 places
```

2.
```
   8.9          89      1 place
×  8.9       ×  89      1 place
   8.01        8 01
  71.2        71 2
  79.21       79.21     2 places
```

3.
```
  6.31         631      2 places
×0.73        ×  73      2 places
  0.1893       1893
  4.417        4 4 17
  4.6063       4.6063   4 places
```

4.
```
  0.29          29      2 places
×0.34        ×  34      2 places
  0.0116       1 16
  0.087        87
  0.0986       0.0986   4 places
```

5. 40 dal
6. 30 L
7. $1.26 - 0.63 = 0.63$
8. $3.052 + 1.98 = 5.032$
9. $400 - 0.5 = 399.5$
10. $6\frac{3}{6} + 8\frac{2}{5} = 6\frac{15}{30} + 8\frac{12}{30} =$
    $14\frac{27}{30} = 14\frac{9}{10}$
11. $3 - 2\frac{1}{4} = 2\frac{4}{4} - 2\frac{1}{4} = \frac{3}{4}$
12. $7\frac{3}{10} + 9\frac{4}{5} = 7\frac{15}{50} + 9\frac{40}{50} =$
    $16\frac{55}{50} = 17\frac{5}{50} = 17\frac{1}{10}$
13. $\frac{3}{\cancel{8}_2} \times \frac{\cancel{4}}{7} = \frac{3}{14}$
14. $\frac{1}{\cancel{2}} \times \frac{\cancel{8}^4}{9} = \frac{4}{9}$
15. $\frac{3}{5} \times \frac{3}{5} = \frac{9}{25}$

SYSTEMATIC REVIEW 10F - LESSON PRACTICE 11C

16. $\frac{2}{3} \times \frac{1}{2} = \frac{1}{3}$

17. $4 \text{ hr} - 1\frac{2}{3} \text{ hr} = 3\frac{3}{3} \text{ hr} - 1\frac{2}{3} \text{ hr} = 2\frac{1}{3} \text{ hours}$

18. $0.2 \text{ gal/mi} \times 56.4 \text{ mi} = 11.28 \text{ gal}$

19. $432 \text{ km} \times 0.6 \text{ mi/km} = 259.2 \text{ mi}$

20. 4,000 meters
    approximately 4,000 yards

Lesson Practice 11A

1. done
2. done
3. $30\% = 0.30$
4. $85\% = 0.85$
5. $10\% = 0.1$
6. $9\% = 0.09$
7. done
8. $5\% = \frac{5}{100} = \frac{1}{20}$
9. $75\% = \frac{75}{100} = \frac{3}{4}$
10. $\frac{1}{1} = 1.00 = 100\%$
11. $\frac{1}{2} = 0.5 = 50\%$
12. $\frac{1}{4} = 0.25 = 25\%$
13. $\frac{3}{4} = 0.75 = 75\%$
14. $\frac{1}{5} = 0.2 = 20\%$
15. $\$3.00 \times 0.06 = \$0.18$
16. $\$7.60 \times 0.15 = \$1.14$
17. $\$7.60 \times 0.50 = \$3.80$ off
    $\$7.60 - \$3.80 = \$3.80$ sale price
18. $\$3.80 \times 0.05 = \$0.19$ tax
    $\$3.80 + \$0.19 = \$3.99$ total

Lesson Practice 11B

1. $38\% = 0.38$
2. $1\% = 0.01$
3. $95\% = 0.95$
4. $71\% = 0.71$
5. $15\% = 0.15$
6. $3\% = 0.03$

7. $10\% = \frac{10}{100} = \frac{1}{10}$
8. $23\% = \frac{23}{100}$
9. $60\% = \frac{60}{100} = \frac{3}{5}$
10. $\frac{1}{4} = 0.25 = 25\%$
11. $\frac{3}{4} = 0.75 = 75\%$
12. $\frac{1}{5} = 0.2 = 20\%$
13. $\frac{1}{1} = 1.00 = 100\%$
14. $\frac{1}{2} = 0.5 = 50\%$
15. $\$7.00 \times 0.05 = \$0.35$
16. $\$10.50 \times 0.16 = \$1.68$
17. $\$12.50 \times 0.20 = \$2.50$ off
    $\$12.50 - \$2.50 = \$10.00$ sale price
    $\$10.00 \times 0.09 = \$0.90$ tax
    $\$10.00 + \$0.90 = \$10.90$ total
18. $0.25 \times 80 = 20$ happy people

Lesson Practice 11C

1. $47\% = 0.47$
2. $5\% = 0.05$
3. $69\% = 0.69$
4. $18\% = 0.18$
5. $32\% = 0.32$
6. $2\% = 0.02$
7. $15\% = \frac{15}{100} = \frac{3}{20}$
8. $17\% = \frac{17}{100}$
9. $50\% = \frac{50}{100} = \frac{1}{2}$
10. $\frac{1}{2} = 0.5 = 50\%$
11. $\frac{1}{5} = 0.2 = 20\%$
12. $\frac{3}{4} = 0.75 = 75\%$

13. $\dfrac{1}{4} = 0.25 = 25\%$

14. $\dfrac{1}{1} = 1.00 = 100\%$

15. done

16. $\$45.00 \times 0.30 = \$13.50$ off
    $\$45.00 - \$13.50 = \$31.50$ sale price
    $\$31.50 \times 0.08 = \$2.52$ tax
    $\$31.50 + \$2.52 = \$34.02$ total

17. $\$18.50 + \$9.50 = \$28.00$ total meals
    $\$28.00 \times 0.2 = \$5.60$ tip
    $\$28.00 + \$5.60 = \$33.60$ total meals and tip

18. $0.50 \times 20 = 10$ people own sleds
    or $\dfrac{1}{2} \times 20 = 10$ people own sleds

Systematic Review 11D

1. done
2. $88\% = 0.88 = \dfrac{88}{100} = \dfrac{22}{25}$
3. $100\% = 1 = \dfrac{100}{100} = 1$
4. $25\% = 0.25 = \dfrac{25}{100} = \dfrac{1}{4}$
5. 50,000 cl
6. 1,200 dm
7. $200 \div 2 = 100$
8. $84 \div 4 = 21$
   $21 \times 3 = 63$
9. $99 \div 3 = 33$
10. 64
11. 1
12. 1,000
13. done
14. $5\dfrac{5}{8} \times 1\dfrac{2}{5} = \dfrac{\overset{9}{45}}{8} \times \dfrac{7}{\underset{1}{5}} = \dfrac{63}{8} = 7\dfrac{7}{8}$
15. $50 \times 0.96 = 48$ questions right
    $50 - 48 = 2$ questions wrong
16. $\$120.00 \times 0.25 = \$30.00$ off
    $\$120.00 - \$30.00 = \$90.00$ sale price
    $\$90.00 \times 0.05 = \$4.50$ tax
    $\$90.00 + \$4.50 = \$94.50$ total

17. 1,000 pieces
    approximately 1 yard

18. $2\dfrac{2}{3} \times 4\dfrac{1}{2} = \dfrac{\overset{4}{8}}{\underset{1}{3}} \times \dfrac{\overset{3}{9}}{\underset{1}{2}} = \dfrac{12}{1} = 12$ miles

Systematic Review 11E

1. $30\% = 0.3 = \dfrac{30}{100} = \dfrac{3}{10}$
2. $72\% = 0.72 = \dfrac{72}{100} = \dfrac{18}{25}$
3. $50\% = 0.5 = \dfrac{50}{100} = \dfrac{1}{2}$
4. $20\% = 0.2 = \dfrac{20}{100} = \dfrac{1}{5}$
5. 50 cl
6. 18,000 ml
7. $120 \div 3 = 40$
8. $72 \div 9 = 8$
   $8 \times 2 = 16$
9. $600 \div 6 = 100$
   $100 \times 5 = 500$
10. 2.448
11. 0.36
12. 0.31
13. 1.26
14. $4\dfrac{2}{7} \times 3\dfrac{1}{5} = \dfrac{\overset{6}{30}}{7} \times \dfrac{16}{\underset{1}{5}} = \dfrac{96}{7} = 13\dfrac{5}{7}$
15. $9\dfrac{1}{5} \times 5\dfrac{4}{5} = \dfrac{46}{5} \times \dfrac{29}{5} = \dfrac{1{,}334}{25} = 53\dfrac{9}{25}$
16. 45% of $20 = 0.45 \times 20 = 9$ boys
17. $\$85.00 \times 0.08 = \$6.80$ shipping and handling
    $\$85.00 + \$6.80 = \$91.80$ total
    $\$91.80 - \$57.00 = \$34.80$ still needed
18. 100 steps
19. $\$9.95 \times 5 = \$49.75$
20. 0.52 lb $+ 0.75$ lb $= 1.27$ lb
    1.27 lb $\times 0.5 = 0.635$ lb

Systematic Review 11F

1. $55\% = 0.55 = \frac{55}{100} = \frac{11}{20}$
2. $40\% = 0.4 = \frac{40}{100} = \frac{2}{5}$
3. $75\% = 0.75 = \frac{75}{100} = \frac{3}{4}$
4. $24\% = 0.24 = \frac{24}{100} = \frac{6}{25}$
5. 2,200 dg
6. 100 dag
7. $6\frac{2}{3} + 1\frac{7}{9} = 6\frac{18}{27} + 1\frac{21}{27} =$
   $7\frac{39}{27} = 8\frac{12}{27} = 8\frac{4}{9}$
8. $7\frac{3}{10} + 9\frac{4}{5} = 7\frac{15}{50} + 9\frac{40}{50} =$
   $16\frac{55}{50} = 17\frac{5}{50} = 17\frac{1}{10}$
9. $3\frac{2}{7} - 1\frac{2}{3} = 3\frac{6}{21} - 1\frac{14}{21} =$
   $2\frac{27}{21} - 1\frac{14}{21} = 1\frac{13}{21}$
10. 0.8547
11. 4.32
12. 1.77
13. 3.33
14. $1\frac{2}{3} \times 4\frac{3}{5} = \frac{5}{3} \times \frac{23}{5} = \frac{23}{3} = 7\frac{2}{3}$
15. $7\frac{3}{5} \times 1\frac{1}{2} = \frac{38}{5} \times \frac{3}{2} = \frac{57}{5} = 11\frac{2}{5}$
16. 85% of 20 = 0.85 × 20 = 17 games won
17. $1.99 \times 0.15 = \$0.2985$
    (rounded to $0.30) off
    $1.99 - \$0.30 = \$1.69$ for one fish
    $1.69 \times 3 = \$5.07$ for three fish
18. 40 km × 0.6 mi/km = 24 mi
19. 8 L × 1.06 qt/L = 8.48 qt
20. $6\frac{2}{3} \times \frac{3}{5} = \frac{20}{3} \times \frac{3}{5} = \frac{4}{1} =$
    $\frac{4}{1} = 4$ bales spoiled

Lesson Practice 12A

1. done
2. $9 = \frac{900}{100} = 900\%$
3. $5 = \frac{500}{100} = 500\%$
4. done
5. $\frac{1}{4} = \frac{25}{100} = 25\%$
6. $\frac{3}{4} = \frac{75}{100} = 75\%$
7. $\frac{1}{5} = \frac{20}{100} = 20\%$
8. $\frac{2}{5} = \frac{40}{100} = 40\%$
9. $\frac{3}{5} = \frac{60}{100} = 60\%$
10. $1\frac{1}{4} = \frac{100}{100} + \frac{25}{100} = \frac{125}{100} = 125\%$
11. $2\frac{1}{2} = \frac{200}{100} + \frac{50}{100} = \frac{250}{100} = 250\%$
12. $3\frac{3}{4} = \frac{300}{100} + \frac{75}{100} = \frac{375}{100} = 375\%$
13. $5\frac{1}{5} = \frac{500}{100} + \frac{20}{100} = \frac{520}{100} = 520\%$
14. done
15. 250% = 2.5
16. 520% = 5.2
17. 100% + 15% + 6% = 121% = 1.21
    $1.21 \times \$18.90 = \$22.869$
    (rounds to $22.87)
18. 400% = 4
    $4 \times \$0.45 = \$1.80$

Lesson Practice 12B

1. $8 = \frac{800}{100} = 800\%$
2. $6 = \frac{600}{100} = 600\%$
3. $2 = \frac{200}{100} = 200\%$
4. $\frac{2}{5} = \frac{40}{100} = 40\%$
5. $\frac{1}{5} = \frac{20}{100} = 20\%$
6. $\frac{4}{5} = \frac{80}{100} = 80\%$
7. $\frac{1}{10} = \frac{10}{100} = 10\%$

8. $\dfrac{9}{10} = \dfrac{90}{100} = 90\%$

9. $\dfrac{1}{4} = \dfrac{25}{100} = 25\%$

10. $3\dfrac{1}{10} = \dfrac{300}{100} + \dfrac{10}{100} = \dfrac{310}{100} = 310\%$

11. $6\dfrac{3}{4} = \dfrac{600}{100} + \dfrac{75}{100} = \dfrac{675}{100} = 675\%$

12. $5\dfrac{1}{2} = \dfrac{500}{100} + \dfrac{50}{100} = \dfrac{550}{100} = 550\%$

13. $2\dfrac{3}{5} = \dfrac{200}{100} + \dfrac{60}{100} = \dfrac{260}{100} = 260\%$

14. $310\% = 3.10$

15. $675\% = 6.75$

16. $100\% = 1$

17. $100\% + 20\% + 5\% = 125\% = 1.25$
    $1.25 \times \$30.00 = \$37.50$

18. $100\% + 5\% = 105\% = 1.05$
    $1.05 \times \$32.95 = \$34.5975$
    (rounds to $34.60)

Lesson Practice 12C

1. $3 = \dfrac{300}{100} = 300\%$

2. $7 = \dfrac{700}{100} = 700\%$

3. $1 = \dfrac{100}{100} = 100\%$

4. $\dfrac{1}{2} = \dfrac{50}{100} = 50\%$

5. $\dfrac{3}{4} = \dfrac{75}{100} = 75\%$

6. $\dfrac{3}{5} = \dfrac{60}{100} = 60\%$

7. $\dfrac{5}{10} = \dfrac{50}{100} = 50\%$

8. $\dfrac{1}{25} = \dfrac{4}{100} = 4\%$

9. $\dfrac{1}{50} = \dfrac{2}{100} = 2\%$

10. $7\dfrac{1}{4} = \dfrac{700}{100} + \dfrac{25}{100} = \dfrac{725}{100} = 725\%$

11. $4\dfrac{1}{5} = \dfrac{400}{100} + \dfrac{20}{100} = \dfrac{420}{100} = 420\%$

12. $1\dfrac{3}{4} = \dfrac{100}{100} + \dfrac{75}{100} = \dfrac{175}{100} = 175\%$

13. $3\dfrac{5}{10} = \dfrac{300}{100} + \dfrac{50}{100} = \dfrac{350}{100} = 350\%$

14. $725\% = 7.25$

15. $420\% = 4.2$

16. $850\% = 8.5$

17. $100\% + 18\% + 2\% = 120\% = 1.2$
    $1.2 \times \$55.20 = \$66.24$

18. $100\% = 1$
    $1 \times 20 = 20$

Systematic Review 12D

1. done

2. $6\dfrac{3}{4} = \dfrac{600}{100} + \dfrac{75}{100} = \dfrac{675}{100} = 6.75 = 675\%$

3. $25\% = 0.25 = \dfrac{25}{100} = \dfrac{1}{4}$

4. $50\% = 0.5 = \dfrac{50}{100} = \dfrac{1}{2}$

5. 600 m

6. 7,000 mm

7. 1,000

8. 100

9. 10

10. $\dfrac{1}{10}$

11. $\dfrac{1}{100}$

12. $\dfrac{1}{1,000}$

13. $\dfrac{3}{4} \div \dfrac{1}{4} = \dfrac{3 \div 1}{1} = 3$

14. $\dfrac{4}{5} \div \dfrac{1}{3} = \dfrac{12}{15} \div \dfrac{5}{15} = \dfrac{12 \div 5}{1} = 2\dfrac{2}{5}$

15. $\dfrac{2}{3} \div \dfrac{1}{4} = \dfrac{8}{12} \div \dfrac{3}{12} = \dfrac{8 \div 3}{1} = \dfrac{8}{3} = 2\dfrac{2}{3}$

16. $100\% + 6\% + 5\% = 111\% = 1.11$
    $1.11 \times \$36.00 = \$39.96$

17. $100\% - 25\% = 75\% = 0.75$
    $0.75 \times \$39.96 = \$29.97$

18. Yes; $\left(10\% = 0.1 = \dfrac{1}{10}\right)$
    $0.1 \times 100 = 10$

Systematic Review 12E

1. $2\dfrac{1}{4} = \dfrac{200}{100} + \dfrac{25}{100} = \dfrac{225}{100} =$
   $225\% = 2.25$
2. $8\dfrac{1}{5} = \dfrac{800}{100} + \dfrac{20}{100} = \dfrac{820}{100} =$
   $820\% = 8.20$
3. $42\% = 0.42 = \dfrac{42}{100} = \dfrac{21}{50}$
4. $150\% = 1.5 = \dfrac{150}{100} = 1\dfrac{50}{100} = 1\dfrac{1}{2}$
5. yard
6. inches
7. quarts
8. ounce
9. pounds
10. miles
11. inch
12. $\dfrac{7}{8} \div \dfrac{1}{3} = \dfrac{21}{24} \div \dfrac{8}{24} = \dfrac{21 \div 8}{1} =$
    $\dfrac{21}{8} = 2\dfrac{5}{8}$
13. $\dfrac{5}{6} \div \dfrac{1}{2} = \dfrac{10}{12} \div \dfrac{6}{12} = \dfrac{10 \div 6}{1} =$
    $\dfrac{10}{6} = 1\dfrac{4}{6} = 1\dfrac{2}{3}$
14. $\dfrac{2}{3} \div \dfrac{1}{4} = \dfrac{8}{12} \div \dfrac{3}{12} = \dfrac{8 \div 3}{1} =$
    $\dfrac{8}{3} = 2\dfrac{2}{3}$
15. $\dfrac{5}{8} \div \dfrac{1}{16} = \dfrac{80}{128} \div \dfrac{8}{128} =$
    $\dfrac{80 \div 8}{1} = \dfrac{10}{1} = 10$ pies
16. $\$235.00 \times 0.05 = \$11.75$ tax
    $\$235.00 + \$11.75 = \$246.75$
17. $100\% - 40\% = 60\% = 0.6$
    $0.6 \times \$12.4 = \$7.44$
18. 68×2.5 cm $= 170$ cm
19. 170 cm $\times 10$ mm/cm $= 1{,}700$ mm
20. Yes; $\left(50\% = \dfrac{1}{2}\right)$
    $\dfrac{1}{2} \times \dfrac{\$40.00}{1} = \dfrac{\$40.00}{2} = \$20.00$
    or
    $0.5 \times \$40.00 = \$20.00$

Systematic Review 12F

1. $7\dfrac{3}{4} = \dfrac{700}{100} + \dfrac{75}{100} = \dfrac{775}{100} =$
   $775\% = 7.75$
2. $5\dfrac{1}{10} = \dfrac{500}{100} + \dfrac{10}{100} = \dfrac{510}{100} =$
   $510\% = 5.10$
3. $9\% = 0.09 = \dfrac{9}{100}$
4. $100\% = 1 = \dfrac{100}{100} = 1$
5. 3 feet or 1 yard
6. 0.4 inch
7. 2.5 cm; 2.54 cm
8. 28 grams
9. 2.2 lb
10. 0.6 miles
11. 1.06 qt
12. $\dfrac{6}{7} \div \dfrac{4}{7} = \dfrac{6 \div 4}{1} = \dfrac{6}{4} = 1\dfrac{2}{4} = 1\dfrac{1}{2}$
13. $\dfrac{3}{5} \div \dfrac{1}{3} = \dfrac{9}{15} \div \dfrac{5}{15} = \dfrac{9 \div 5}{1} = \dfrac{9}{5} = 1\dfrac{4}{5}$
14. $\dfrac{5}{8} \div \dfrac{2}{3} = \dfrac{15}{24} \div \dfrac{16}{24} = \dfrac{15 \div 16}{1} = \dfrac{15}{16}$
15. $\dfrac{4}{5} \div \dfrac{2}{5} = \dfrac{4 \div 2}{1} = \dfrac{4}{2} = 2$ friends
16. $20 \times 3 = 60$ balloons
17. $100\% + 20\% + 4\% = 124\% = 1.24$
    $1.24 \times \$15.00 = \$18.60$
18. 20 oz $\times 28$ g/oz $= 560$ g
19. 560 g $\times 1{,}000$ mg/g $= 560{,}000$ mg
20. $100\% - 20\% = 80\% = 0.8$
    $0.8 \times \$355.00 = \$284.00$

Lesson Practice 13A

1. brown
2. $0.5 \times 20 = 10$ people have brown hair
3. 5   (There are the same number with blond and red hair.)
4. vanilla
5. strawberry
6. $300 \times 0.3 = 90$ chocolate cones

7. grew

 did not grow

(The position of the colored sections is not important as long as the right number of sections are colored.)

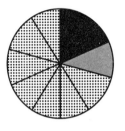

8. giving

 savings

 other

Lesson Practice 13B

1. sleep
2. $0.25 \times 24 \text{ hours} = 6 \text{ hours}$
3. $25\% = \frac{25}{100} = \frac{1}{4}$
4. repairs
5. gasoline
6. $\$1,000.00 \times 0.3 = \$300.00$
7. cats only

     dogs only

     both

    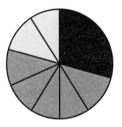

8. ice cream

     pie

     cake

    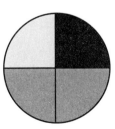

Lesson Practice 13C

1. red, white, and blue
2. blue and white
3. $600 \times 0.35 = 210$ red cars
4. pine
5. $480 \times 0.4 = 192$ pine trees
6. $480 \times 0.3 = 144$ spruce trees
7. goldfish

     minnows

    guppies

    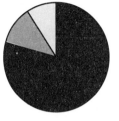

8. wood

     brick

     stone

    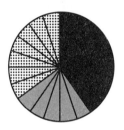

Systematic Review 13D

1. corn
2. $10\% + 15\% = 25\% = \frac{25}{100} = \frac{1}{4}$
3. $500 \times 0.75 = 375$
4. $25\% = 0.25 = \frac{25}{100} = \frac{1}{4}$
5. $40\% = 0.4 = \frac{40}{100} = \frac{2}{5}$
6. $300\% = 3 = \frac{300}{100} = 3$
7. $15\% = 0.15 = \frac{15}{100} = \frac{3}{20}$
8. $4\frac{1}{2} = \frac{400}{100} + \frac{50}{100} = \frac{450}{100} = 450\% = 4.5$
9. 2.2 lb
10. 0.6 miles
11. 1.06 qt
12. $1.79 \times 0.3 = 0.537$
13. $25 - 0.45 = 24.55$
14. $0.135 + 0.4 = 0.535$

SYSTEMATIC REVIEW 13D - SYSTEMATIC REVIEW 13F

15. $4\frac{2}{3} \div 1\frac{1}{5} = \frac{14}{3} \div \frac{6}{5} = \frac{70}{15} \div \frac{18}{15} =$
    $\frac{70 \div 18}{1} = \frac{70}{18} = \frac{35}{9} = 3\frac{8}{9}$

16. $4\frac{1}{8} \div 2\frac{2}{3} = \frac{33}{8} \div \frac{8}{3} = \frac{99}{24} \div \frac{64}{24} =$
    $\frac{99 \div 64}{1} = \frac{99}{64} = 1\frac{35}{64}$

17. $7\frac{5}{10} \div 1\frac{1}{2} = \frac{75}{10} \div \frac{3}{2} = \frac{150}{20} \div \frac{30}{20} =$
    $\frac{150 \div 30}{1} = \frac{150}{30} = 5$ bags

18. $100\% + 7\% = 107\% = 1.07$
    $\$35.00 \times 1.07 = \$37.45$

Systematic Review 13E

1. biking
    running
   ☐ swimming
   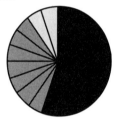

2. 50 miles $\times 0.1 = 5$ miles
3. $50\% = 0.5 = \frac{50}{100} = \frac{1}{2}$
4. $20\% = 0.2 = \frac{20}{100} = \frac{1}{5}$
5. $5\% = 0.05 = \frac{5}{100} = \frac{1}{20}$
6. $700\% = 7 = \frac{700}{100} = 7$
7. $6\frac{1}{4} = \frac{600}{100} + \frac{25}{100} =$
   $\frac{625}{100} = 625\% = 6.25$
8. 0.4 inches
9. 2.5 cm (or 2.54 cm)
10. 28 grams
11. $2.45 \times 0.09 = .2205$
12. $0.7 - 0.15 = 0.55$
13. $1.58 + 7.6 = 9.18$
14. $3\frac{1}{6} \div 1\frac{1}{5} = \frac{19}{6} \div \frac{6}{5} = \frac{95}{30} \div \frac{36}{30}$
    $= \frac{95 \div 36}{1} = \frac{95}{36} = 2\frac{23}{36}$

15. $1\frac{1}{4} \div 1\frac{1}{8} = \frac{5}{4} \div \frac{9}{8} = \frac{40}{32} \div \frac{36}{32}$
    $= \frac{40 \div 36}{1} = \frac{40}{36} = \frac{10}{9} = 1\frac{1}{9}$

16. $2\frac{1}{6} \div 1\frac{4}{5} = \frac{13}{6} \div \frac{9}{5} = \frac{65}{30} \div \frac{54}{30}$
    $= \frac{65 \div 54}{1} = \frac{65}{54} = 1\frac{11}{54}$

17. $6\frac{3}{4} \div 1\frac{1}{4} = \frac{27}{4} \div \frac{5}{4} = \frac{27 \div 5}{1}$
    $= \frac{27}{5} = 5\frac{2}{5}$

    Five people can be served. There will be pizza left over because she can't serve pizza to a part of a person.

18. $100\% + 10\% = 110\% = 1.1$
    $\$1.30 \times 1.1 = \$1.43$
19. 55.6 mi/hr $\times 7$ hr $= 389.2$ miles
20. 2 kg $\times 2.2$ lb/kg $= 4.4$ lb

Systematic Review 13F

1. clouds
2. $30 \times 0.2 = 6$ days
3. $30 \times 0.1 = 3$ days
4. $75\% = 0.75 = \frac{75}{100} = \frac{3}{4}$
5. $2\% = 0.02 = \frac{2}{100} = \frac{1}{50}$
6. $500\% = 5 = \frac{500}{100} = 5$
7. $30\% = 0.3 = \frac{30}{100} = \frac{3}{10}$
8. $3\frac{1}{5} = \frac{300}{100} + \frac{20}{100} = \frac{320}{100} = 320\% = 3.2$
9. 6,000 g
10. 25,000 ml
11. $0.17 \times 0.58 = 0.0986$
12. $1.75 - 0.9 = 0.85$
13. $3.68 + 0.061 = 3.741$
14. $3\frac{1}{2} \div 2\frac{2}{5} = \frac{7}{2} \div \frac{12}{5} = \frac{35}{10} \div \frac{24}{10} =$
    $\frac{35 \div 24}{1} = \frac{35}{24} = 1\frac{11}{24}$

15. $5\frac{1}{2} \div 2\frac{2}{3} = \frac{11}{2} \div \frac{8}{3} = \frac{33}{6} \div \frac{16}{6} =$
    $\frac{33 \div 16}{1} = \frac{33}{16} = 2\frac{1}{16}$

16. $3\frac{1}{5} \div 1\frac{3}{4} = \frac{16}{5} \div \frac{7}{4} = \frac{64}{20} \div \frac{35}{20} =$
    $\frac{64 \div 35}{1} = \frac{64}{35} = 1\frac{29}{35}$

17. 200 cm
    200 cm × 0.4 in/cm = 80 in

18. 60 × 1.2 = 72 doughnuts

19. 1.8 cups/hr × 8 hr = 14.4 cups

20. 50 km × 0.6 mi/km = 30 mi

Lesson Practice 14A

1. done
2. done
3. Estimate:

   20.
   × 0.1
   2.0

   22.163   22163   3 places
   × 0.11   × 11   2 places
   0.22163   22163
   2.2163   2 2163
   2.43793   2.43793 5 places

4. Estimate:

   8
   × 2
   16

   8.246   8246   3 places
   × 2.159   × 2 159 3 places
   0.074214   7 42 14
   0.41230   4 1230
   0.8246   8246
   16.492   16 492
   17.803114   17.803 114 6 places

5. Estimate:

   100.
   × 0.6
   60.

   135.004   135004   3 places
   × 0.58   × 58   2 places
   10.80032   10 80032
   67.5020   67 5020
   78.30232   78.30232 5 places

6. Estimate:

   40.
   × 0.6
   24.

   41.685   4 1685   3 places
   × 0.555   × 555   3 places
   0.208425   208425
   2.08425   208425
   20.8425   208425
   23.135175   23.135175 6 places

7. 9.625 mi/hr × 0.25 hr = 2.40625 mi

8. 12.75 × $1.539 = $19.62225
   (rounded to $19.62)

Lesson Practice 14B

1. Estimate:

   50.
   × 0.005
   0.25

   50.345   50345   3 places
   × 0.005   × 005   3 places
   0.251725   0.251725 6 places

2. Estimate:

   31
   × 2
   62

   31.489   3 1489   3 places
   × 2.23   × 223   2 places
   0.94467   94467
   6.2978   6 2978
   62.978   62 978
   70.22047   70.22047 5 places

LESSON PRACTICE 14B - LESSON PRACTICE 14C

3. Estimate:

   0.9
   ×0.009
   0.0081

   0.876   876   3 places
   ×0.009   ×   9   3 places
   0.007884   0.007884 6 places

4. Estimate:

   6
   ×1
   6

   6.218   6 218   3 places
   ×1.064   × 1 064   3 places
   0.024872   24 872
   0.37308   37 308
   6.218   6 218
   6.615952   6.615952   6 places

5. Estimate:

   500.
   × 0.2
   100

   450.132   450 132   3 places
   ×   0.222   ×   222   3 places
   0.900264   900 264
   9.00264   9 00 264
   90.0264   90 0 264
   99.929304   99.929304 6 places

6. Estimate:

   2
   ×0.8
   1.6

   1.539   1539   3 places
   ×0.82   ×   82   2 places
   0.03078   3078
   1.2312   1 231 2
   1.26198   1.26 198   5 places

7. 39.37 in/m × 0.001 m = 0.03937 in
8. $1.279 × 1.30 = $1.6627
   (rounded to $1.663)

Lesson Practice 14C

1. Estimate:

   60
   × 0.03
   1.80

   61.198   6 1198   3 places
   × 0.03   ×   3   2 places
   1.83594   1.83594   5 places

2. Estimate:

   40
   × 1
   40

   42.345   42 345   3 places
   × 1.16   ×   116   2 places
   2.54070   2 54070
   4.2345   4 2345
   42.345   42.345
   49.12020   49.12020 5 places

3. Estimate:

   0.5
   ×0.8
   0.40

   0.511   511   3 places
   ×0.8   ×   8   1 place
   0.4088   0.4088   4 places

4. Estimate:

   8
   × 3
   24

   7.639   7639   3 places
   × 2.804   × 2804   3 places
   0.030556   30 556
   6.1112   6 11120
   15.278   15 278
   21.419756   21.419756 6 places

LESSON PRACTICE 14C - SYSTEMATIC REVIEW 14E

5. Estimate:

$$\begin{array}{r} 20 \\ \times\ 0.1 \\ \hline 2 \end{array}$$

$$\begin{array}{r} 20.352 \\ \times\ 0.136 \\ \hline 0.122112 \\ 0.61056 \\ 2.0352 \\ \hline 2.767872 \end{array}$$

$$\begin{array}{r} 20352 \quad \text{3 places} \\ \times\ \quad 136 \quad \text{3 places} \\ \hline 122112 \\ 61056 \\ 2\,0352 \\ \hline 2.767872 \quad \text{6 places} \end{array}$$

6. Estimate:

$$\begin{array}{r} 9 \\ \times 0.7 \\ \hline 6.3 \end{array}$$

$$\begin{array}{r} 9.293 \\ \times 0.71 \\ \hline 0.09293 \\ 6.5051 \\ \hline 6.59803 \end{array}$$

$$\begin{array}{r} 9293 \quad \text{3 places} \\ \times\ \quad 71 \quad \text{2 places} \\ \hline 9293 \\ 6\,5051 \\ \hline 6.59803 \quad \text{5 places} \end{array}$$

$$\begin{array}{r} 9293 \quad \text{3 places} \\ \times 71 \quad +2 \text{ places} \\ \hline 9293 \\ 65051 \\ \hline 6.59803 \quad \text{5 places} \end{array}$$

7. $1.06 \text{ qt/L} \times 0.001 \text{ L} = 0.00106 \text{ qt}$

8. $1.667 \text{ gal/min} \times 60 \text{ min} = 100.02 \text{ gal}$

Systematic Review 14D

1. Estimate:

$$\begin{array}{r} 2 \\ \times 0.6 \\ \hline 1.2 \end{array} \qquad \begin{array}{r} 1724 \quad \text{3 places} \\ \times \quad 6 \quad \text{1 place} \\ \hline 1.0344 \quad \text{4 places} \end{array}$$

2. Estimate:

$$\begin{array}{r} 3 \\ \times 0.4 \\ \hline 1.2 \end{array} \qquad \begin{array}{r} 3465 \quad \text{3 places} \\ \times \quad 403 \quad \text{3 places} \\ \hline 10395 \\ 13860 \\ \hline 1.396395 \quad \text{6 places} \end{array}$$

3. $50\% = 0.5 = \frac{50}{100} = \frac{1}{2}$

4. $8\% = 0.08 = \frac{8}{100} = \frac{2}{25}$

5. $600\% = 6 = \frac{600}{100} = 6$

6. $20\% = 0.2 = \frac{20}{100} = \frac{1}{5}$

7. $5\frac{1}{4} = \frac{500}{100} + \frac{25}{100} = \frac{525}{100} = 525\% = 5.25$

8. 70,000 cm

9. 100 mg

10. $6\frac{1}{8} + 3\frac{1}{2} = 6\frac{2}{16} + 3\frac{8}{16} = 9\frac{10}{16} = 9\frac{5}{8}$

11. $10\frac{2}{5} - 4\frac{5}{8} = 10\frac{16}{40} - 4\frac{25}{40} =$
    $9\frac{56}{40} - 4\frac{25}{40} = 5\frac{31}{40}$

12. $\frac{\cancel{7}}{\cancel{3}} \times \frac{\cancel{6}^{2}}{\cancel{7}} = \frac{2}{1} = 2$

13. $\frac{3}{5} \times \frac{2}{1} = \frac{6}{5} = 1\frac{1}{5}$

14. $\frac{7}{\cancel{8}_{2}} \times \frac{\cancel{4}}{3} = \frac{7}{6} = 1\frac{1}{6}$

15. $\frac{3}{8} \times \frac{5}{1} = \frac{15}{8} = 1\frac{7}{8}$

16. $\frac{10}{16} \div \frac{1}{8} = \frac{10}{\cancel{16}_{2}} \times \frac{\cancel{8}}{1} = \frac{10}{2} = 5$ pieces

17. $\$15.50 \times 1.06 = \$16.43$

18. $0.6 \text{ mi/km} \times 0.001 \text{ km} = 0.0006 \text{ mi}$

Systematic Review 14E

1. Estimate:

$$\begin{array}{r} 2 \\ \times 0.007 \\ \hline 0.014 \end{array} \qquad \begin{array}{r} 2192 \quad \text{3 places} \\ \times \quad 7 \quad \text{3 places} \\ \hline 0.015344 \quad \text{6 places} \end{array}$$

2. Estimate:

$$\begin{array}{r} 10 \\ \times\ 0.8 \\ \hline 8 \end{array} \qquad \begin{array}{r} 13543 \quad \text{3 places} \\ \times \quad 75 \quad \text{2 places} \\ \hline 67715 \\ 94801 \\ \hline 10.15725 \quad \text{5 places} \end{array}$$

3. summer

4. spring

5. $850 \times 0.10 = 85$

6. $25\% = 0.25 = \frac{25}{100} = \frac{1}{4}$

7. $100\% = 1 = \frac{100}{100} = 1$

SYSTEMATIC REVIEW 14E - LESSON PRACTICE 15A

8. $250\% = 2.5 = \frac{250}{100} = 2\frac{1}{2}$

9. $16\% = 0.16 = \frac{16}{100} = \frac{4}{25}$

10. $1\frac{3}{4} = \frac{100}{100} + \frac{75}{100} = \frac{175}{100} = 175\% = 1.75$

11. 12,000 L

12. 300 cm

13. $\frac{5}{{}_3\cancel{9}} \times \frac{\cancel{3}}{2} = \frac{5}{6}$

14. $\frac{9}{{}_2\cancel{10}} \times \frac{\cancel{5}}{1} = \frac{9}{2} = 4\frac{1}{2}$

15. $\frac{1}{3} \times \frac{4}{3} = \frac{4}{9}$

16. $8 \text{ hr} \times 0.08 = 0.64$ hours

17. $15.25 \times \$1.549 =$
    $\$23.62225$ (rounded $\$23.62$)

18. $1.56 \text{ lb} + 2.13 \text{ lb} + 1.5 \text{ lb} + 0.66 \text{ lb} = 5.85 \text{ lb}$

19. $5.85 \text{ lb} \times 16 \text{ oz/lb} = 93.6 \text{ oz}$
    $93.6 \text{ oz} \times 28 \text{ g/oz} = 2,620.8 \text{ g}$

20. $0.25 \times 12 = 3$ oranges to her mother
    $\frac{1}{4} \times \frac{12}{1} = \frac{12}{4} = 3$ oranges to her sister
    $12 - 6 = 6$ left
    $6 \div 12 = 0.5 = 50\%$ left

Systematic Review 14F

1. Estimate:

   $\begin{array}{r} 0.6 \\ \times 0.1 \\ \hline 0.06 \end{array}$

   $\begin{array}{r} 567 \quad \text{3 places} \\ \times \quad 148 \quad \text{3 places} \\ \hline 4536 \\ 2268 \\ 567 \\ \hline 0.083916 \quad \text{6 places} \end{array}$

2. Estimate:

   $\begin{array}{r} 2 \\ \times 1 \\ \hline 2 \end{array}$

   $\begin{array}{r} 1905 \quad \text{3 places} \\ \times \quad 1321 \quad \text{3 places} \\ \hline 1905 \\ 3810 \\ 5715 \\ 1905 \\ \hline 2.516505 \quad \text{6 places} \end{array}$

3. oak
    maple
   hemlock

4. $600 \times 0.05 = 30$ hemlock trees

5. $75\% = 0.75 = \frac{75}{100} = \frac{3}{4}$

6. $45\% = 0.45 = \frac{45}{100} = \frac{9}{20}$

7. $125\% = 1.25 = \frac{125}{100} = 1\frac{1}{4}$

8. $1\% = 0.01 = \frac{1}{100}$

9. $7\frac{1}{5} = \frac{700}{100} + \frac{20}{100} = \frac{720}{100} = 720\% = 7.2$

10. 35,000 mg

11. 1,100 dam

12. 500 cm

13. $\frac{2}{5} \times \frac{8}{3} = \frac{16}{15} = 1\frac{1}{15}$

14. $\frac{\cancel{4}^2}{5} \times \frac{3}{\cancel{2}} = \frac{6}{5} = 1\frac{1}{5}$

15. $\frac{\cancel{3}}{\cancel{4}} \times \frac{\cancel{4}}{\cancel{3}} = 1$

16. $900 \times 0.15 = 135$ yellow roses

17. $\$9.00 \times 1.10 = \$9.90$

18. $90 \text{ cm} \times 0.4 \text{ in/cm} = 36 \text{ in}$

19. $5.725 \text{ tons} \times 0.1 = 0.5725 \text{ ton}$

20. $0.5725 \text{ ton} \times 2,000 \text{ lb/ton} = 1,145 \text{ lb}$

Lesson Practice 15A

1. done

2. done

3. $750 \text{ d}\cancel{\text{g}} \times \frac{1 \cancel{\text{g}}}{10 \text{ d}\cancel{\text{g}}} \times \frac{1 \text{ dag}}{10 \cancel{\text{g}}} = 7.5 \text{ dag};$
   two places greater;
   $750 \text{ dg} = 7.5 \text{ dag}$

4. 0.014 km

5. 0.4 km

6. 5.3 L

7. 1.3 cg

LESSON PRACTICE 15A - SYSTEMATIC REVIEW 15D

8. 0.48 dam
9. 0.07 hl
10. 0.000609 kg
11. 0.5 cm
12. 1.2 dal

Lesson Practice 15B
1. 30 L
2. 5 m
3. 10 kl
4. 0.25 km
5. 0.18 g
6. 0.45 kl
7. 0.5 km
8. 80 g
9. 8.3 cg
10. 0.008 kl
11. 1.02 m
12. 0.01 km

Lesson Practice 15C
1. 0.015 dam
2. 8 g
3. 1 hm
4. 0.25 km
5. 30 g
6. 0.034 dag
7. 0.99 m
8. 7.3 dam
9. 15 kl
10. 50 m
11. 45.638 g
12. 300 kg

Systematic Review 15D
1. 0.45 g
2. 60 cm
3. 100 mg

4. Estimate:
$$\begin{array}{r} 0.6 \\ \times 0.5 \\ \hline 0.30 \end{array}$$

$$\begin{array}{r} 621 \text{ 3 places} \\ \times 543 \text{ 3 places} \\ \hline 1863 \\ 2484 \\ 3105 \\ \hline 0.337203 \text{ 6 places} \end{array}$$

5. Estimate:
$$\begin{array}{r} 3 \\ \times 2 \\ \hline 6 \end{array}$$

$$\begin{array}{r} 27.28 \text{ 3 places} \\ \times 1.69 \text{ 2 places} \\ \hline 24552 \\ 16368 \\ 2728 \\ \hline 4.61032 \text{ 5 places} \end{array}$$

6. $11\% = 0.11 = \frac{11}{100}$

7. $5\% = 0.05 = \frac{5}{100} = \frac{1}{20}$

8. $100\% = 1 = \frac{100}{100} = 1$

9. $25\% = 0.25 = \frac{25}{100} = \frac{1}{4}$

10. $18 \div 2 = 9$
    $9 \times 1 = 9$

11. $69 \div 3 = 23$
    $23 \times 2 = 46$

12. $248 \div 8 = 31$
    $31 \times 5 = 155$

13. $4 \div 4 = 1$

14. $7\frac{7}{8} \div 1\frac{1}{2} = \frac{63}{8} \div \frac{3}{2} =$
    $\frac{\cancel{63}^{21}}{\cancel{8}_4} \times \frac{\cancel{2}}{\cancel{3}} = \frac{21}{4} = 5\frac{1}{4}$

15. $3\frac{2}{7} \div 2\frac{1}{3} = \frac{23}{7} \div \frac{7}{3} =$
    $\frac{23}{7} \times \frac{3}{7} = \frac{69}{49} = 1\frac{20}{49}$

16. $24 \text{ strides} \times \frac{1 \text{ m}}{1 \text{ stride}} \times \frac{1 \text{ dam}}{10 \text{ m}} = 2.4 \text{ dam}$

17. $105 \text{ kg} \times 2.2 \text{ lb/kg} = 231 \text{ lb}$

18. $\$1.679 \times 2 = \$3.358$

Systematic Review 15E

1. 0.39 m
2. 3,800 g
3. 0.087 kl
4. Estimate:

   $$\begin{array}{r} 50 \\ \times\ 0.6 \\ \hline 30 \end{array} \qquad \begin{array}{r} 48003 \quad \text{3 places} \\ \times\ 61 \quad \text{2 places} \\ \hline 48003 \\ 2\,88\,0\,18 \\ \hline 29.28183 \quad \text{5 places} \end{array}$$

5. Estimate:

   $$\begin{array}{r} 20 \\ \times\ 30 \\ \hline 600 \end{array} \qquad \begin{array}{r} 2277 \quad \text{2 places} \\ \times\ 3154 \quad \text{2 places} \\ \hline 9108 \\ 1\,1385 \\ 2277 \\ 6831 \\ \hline 718.1658 \quad \text{4 places} \end{array}$$

6. $250\% = 2.5 = \dfrac{250}{100} = 2\dfrac{50}{100} = 2\dfrac{1}{2}$
7. $34\% = 0.34 = \dfrac{34}{100} = \dfrac{17}{50}$
8. $7\% = 0.07 = \dfrac{7}{100}$
9. $75\% = 0.75 = \dfrac{75}{100} = \dfrac{3}{4}$
10. $400 \div 10 = 40$
    $40 \times 1 = 40$
11. $192 \div 6 = 32$
    $32 \times 5 = 160$
12. $96 \div 4 = 24$
    $24 \times 3 = 72$
13. $100 \div 5 = 20$
    $20 \times 4 = 80$
14. $6\dfrac{1}{3} \div 2\dfrac{1}{8} = \dfrac{19}{3} \div \dfrac{17}{8} =$
    $\dfrac{19}{3} \times \dfrac{8}{17} = \dfrac{152}{51} = 2\dfrac{50}{51}$
15. $8\dfrac{8}{10} \div 2\dfrac{2}{3} = \dfrac{88}{10} \div \dfrac{8}{3} =$
    $\dfrac{\overset{11}{\cancel{88}}}{10} \times \dfrac{3}{\cancel{8}} = \dfrac{33}{10} = 3\dfrac{3}{10}$
16. 61 km
17. 110 mg
18. $\$79.97 \times 0.37 = \$29.5889$ off
    (rounded to $\$29.59$)
    $\$79.97 - \$29.59 = \$50.38$ sale price
19. $\$8.95 + \$8.95 + \$12.49 = \$30.39$
    $\$30.39 \times 1.16 = \$35.2524$ total
    (rounded to $\$35.25$)
20. $2\dfrac{2}{5}\text{ mi} \times 1\dfrac{2}{3}\text{ mi} = \dfrac{\overset{4}{\cancel{12}}}{\cancel{5}} \times \dfrac{\cancel{5}}{\cancel{3}} = 4 \text{ mi}$

Systematic Review 15F

1. 0.005 g
2. 170,000 cm
3. 0.5 hl
4. Estimate:

   $$\begin{array}{r} 3 \\ \times 0.006 \\ \hline 0.018 \end{array} \qquad \begin{array}{r} 2952 \quad \text{3 places} \\ \times\ 6 \quad \text{3 places} \\ \hline 0.017712 \quad \text{6 places} \end{array}$$

5. Estimate:

   $$\begin{array}{r} 100 \\ \times\ 0.6 \\ \hline 60 \end{array} \qquad \begin{array}{r} 135123 \quad \text{3 places} \\ \times\ 612 \quad \text{3 places} \\ \hline 270246 \\ 1\,35123 \\ 81\,0738 \\ \hline 82.695276 \quad \text{6 places} \end{array}$$

6. $2\% = 0.02 = \dfrac{2}{100} = \dfrac{1}{50}$
7. $20\% = 0.2 = \dfrac{20}{100} = \dfrac{1}{5}$
8. $60\% = 0.6 = \dfrac{60}{100} = \dfrac{3}{5}$
9. $900\% = 9 = \dfrac{900}{100} = 9$
10. $54 \div 6 = 9$
    $9 \times 1 = 9$
11. $88 \div 8 = 11$
    $11 \times 3 = 33$
12. $490 \div 7 = 70$
    $70 \times 5 = 350$
13. $300 \div 5 = 60$
    $60 \times 2 = 120$

14. $5\frac{1}{9} \div 2\frac{1}{6} = \frac{46}{9} \div \frac{13}{6} =$

$\frac{46}{\cancel{9}3} \times \frac{\cancel{6}^2}{13} = \frac{92}{39} = 2\frac{14}{39}$

15. $7\frac{1}{3} \div \frac{2}{5} = \frac{22}{3} \div \frac{2}{5} =$

$\frac{\cancel{22}^{11}}{3} \times \frac{5}{\cancel{2}} = \frac{55}{3} = 18\frac{1}{3}$

16. $8\frac{3}{4} \text{ lb} \div 5 = \frac{35}{4} \text{ lb} \div \frac{5}{1} =$

$\frac{\cancel{35}^7}{4} \text{ lb} \times \frac{1}{\cancel{5}} = \frac{7}{4} \text{ lb} = 1\frac{3}{4} \text{ lb}$

17. 0.5 kl
18. $68 \times 2.54 \text{ cm} = 172.72 \text{ cm} = 1,727.2 \text{ mm}$
19. $\$48.00 \times 1.25 = \$60.00$ from Dad
$\$48.00 + \$60.00 = \$108.00$ total
20. $0.2 \times 100 = 20$ days

$\frac{1}{5} \times \frac{100}{1} = \frac{100}{5} = 20$ days

Lesson Practice 16A

1. done
2. done
3. $A \approx 3.14(1.6 \text{ m})^2 = 8.0384 \text{ m}^2$
$C \approx 2(3.14)(1.6 \text{ m}) = 10.048 \text{ m}$
4. $A \approx 3.14(15 \text{ ft})^2 = 706.5 \text{ ft}^2$
$C \approx (3.14)(30 \text{ ft}) = 94.2 \text{ ft}$
5. $A \approx 3.14(12 \text{ ft})^2 = 452.16 \text{ ft}^2$
6. $C \approx 2(3.14)(7.5 \text{ ft}) = 47.1 \text{ ft}$
7. $C \approx (3.14)(12 \text{ in}) = 37.68 \text{ in}$
37.68" > 36"; not enough
8. $A \approx 3.14(9 \text{ m})^2 = 254.34 \text{ m}^2$

Lesson Practice 16B

1. $A \approx 3.14(3 \text{ in})^2 = 28.26 \text{ in}^2$
$C \approx 2(3.14)(3 \text{ in}) = 18.84 \text{ in}$
2. $A \approx 3.14(5 \text{ ft})^2 = 78.5 \text{ ft}^2$
$C \approx (3.14)(10 \text{ ft}) = 31.4 \text{ ft}$
3. $A \approx 3.14(4.2 \text{ m})^2 = 55.3896 \text{ m}^2$
$C \approx 2(3.14)(4.2 \text{ m}) = 26.376 \text{ m}$

4. $A \approx 3.14(6 \text{ ft})^2 = 113.04 \text{ ft}^2$
$C \approx (3.14)(12 \text{ ft}) = 37.68 \text{ ft}$
5. $A \approx 3.14(16 \text{ ft})^2 = 803.84 \text{ ft}^2$
6. $C \approx 2(3.14)(11 \text{ ft}) = 69.08 \text{ ft}$
7. $C \approx (3.14)(5.4 \text{ mm}) = 16.956 \text{ mm}$
8. $A \approx 3.14(8.5 \text{ ft})^2 = 226.865 \text{ ft}^2$

Lesson Practice 16C

1. $A \approx 3.14(6 \text{ in})^2 = 113.04 \text{ in}^2$
$C \approx 2(3.14)(6 \text{ in}) = 37.68 \text{ in}$
2. $A \approx 3.14(7 \text{ ft})^2 = 153.86 \text{ ft}^2$
$C \approx (3.14)(14 \text{ ft}) = 43.96 \text{ ft}$
3. $A \approx 3.14(10 \text{ m})^2 = 314 \text{ m}^2$
$C \approx 2(3.14)(10 \text{ m}) = 62.8 \text{ m}$
4. $A \approx 3.14(4.75 \text{ ft})^2 = 70.84625 \text{ ft}^2$
$C \approx (3.14)(9.5 \text{ ft}) = 29.83 \text{ ft}$
5. $A \approx 3.14(6 \text{ in})^2 = 113.04 \text{ in}^2$
(rounded to 113 in^2)
6. $A \approx 3.14(9 \text{ ft})^2 = 254.34 \text{ ft}^2$
254.34 ft > 200 ft; not enough
7. $C \approx 2(3.14)(9 \text{ ft}) = 56.52 \text{ ft}$
56.52 ft ÷ 2 = 28.26 pieces;
29 pieces must be bought.
8. $C \approx 2(3.14)(100 \text{ mi}) = 628$ miles

Systematic Review 16D

1. $A \approx 3.14(1.4 \text{ in})^2 = 6.1544 \text{ in}^2$
$C \approx 2(3.14)(1.4 \text{ in}) = 8.792 \text{ in}$
2. $A \approx 3.14(10 \text{ ft})^2 = 314 \text{ ft}^2$
$C \approx 3.14(20 \text{ ft}) = 62.8 \text{ ft}$
3. 0.15 m
4. 0.8 kl
5. 3,500 cg
6. $1.65 \times 9.43 = 15.5595$
7. $0.209 \times 8.07 = 1.68663$
8. $5.061 \times 3.94 = 19.94034$

9. $6\frac{1}{5} \div 1\frac{3}{4} = \frac{31}{5} \div \frac{7}{4} = \frac{31}{5} \times \frac{4}{7} =$
$\frac{124}{35} = 3\frac{19}{35}$

10. $4\frac{3}{5} \div 4\frac{1}{2} = \frac{23}{5} \div \frac{9}{2} = \frac{23}{5} \times \frac{2}{9} =$
$\frac{46}{45} = 1\frac{1}{45}$

11. $\frac{1}{2} \div \frac{1}{8} = \frac{1}{2} \times \frac{8}{1} = 4$

12. done

13. $10' + 10' + 10' + 10' = 40'$

14. $7\ m + 15\ m + 7\ m + 15\ m = 44\ m$

15. $12\ m + 13\ m + 12\ m + 13\ m = 50\ m$

16. $13\ m + 12\ m + 13\ m = 38\ m$

17. $38 \times 3\ ft = 114\ ft$ (approximately)

18. $\$18,550.00 \times 0.02 = \$371.00$

Systematic Review 16E

1. $A = 3.14(3.2\ cm)^2 = 32.1536\ cm^2$
$C = 2(3.14)(3.2\ cm) = 20.096\ cm$

2. $A = 3.14(2\ m)^2 = 12.56\ m^2$
$C = 3.14(4\ m) = 12.56\ m$

3. $0.01\ km$

4. $5\ cl$

5. $20,000\ dg$

6. $0.2 \times 0.32 = 0.064$

7. $1.45 \times 0.04 = 0.058$

8. $0.005 \times 0.02 = 0.0001$

9. $3\frac{1}{2} \div 2\frac{1}{6} = \frac{7}{2} \div \frac{13}{6} =$
$\frac{7}{2} \times \frac{6}{13} = \frac{21}{13} = 1\frac{8}{13}$

10. $5\frac{1}{4} \div 3\frac{1}{4} = \frac{21}{4} \div \frac{13}{4} =$
$\frac{21}{4} \times \frac{4}{13} = \frac{21}{13} = 1\frac{8}{13}$

11. $\frac{7}{8} \div \frac{1}{16} = \frac{7}{8} \times \frac{16}{1} = 14$

12. $4" + 10" + 4" + 10" = 28"$

13. $45' + 45' + 45' + 45' = 180'$

14. $2.3\ m + 1.2\ m + 2.3\ m + 1.2\ m = 7\ m$

15. $\$2,000.00 \times 0.23 = \$460.00$

16. $3.14(7\ in)^2 = 153.86\ in^2$
(round to $154\ in^2$)

17. $50\ m + 75\ m + 50\ m + 75\ m = 250\ m$

18. $\$12.35/m \times 250\ m = \$3,087.50$

19. $4\ kl$

20. $3\frac{1}{2} \div \frac{1}{4} = \frac{7}{2} \div \frac{1}{4} =$
$\frac{7}{2} \times \frac{4}{1} = 14$ containers

Systematic Review 16F

1. $A \approx 3.14(1\ m)^2 = 3.14\ m^2$
$C \approx 2(3.14)(1\ m) = 6.28\ m$

2. $A \approx 3.14(0.5\ km)^2 = 0.785\ km^2$
$C \approx 3.14(1\ km) = 3.14\ km$

3. $1.7\ cm$

4. $0.32\ L$

5. $5,000\ g$

6. $0.06 \times 0.05 = 0.003$

7. $45.1 \times 0.8 = 36.08$

8. $116 \times 0.19 = 22.04$

9. $7\frac{1}{8} \div 3\frac{3}{4} = \frac{57}{8} \div \frac{15}{4} =$
$\frac{57}{8} \times \frac{4}{15} = \frac{19}{10} = 1\frac{9}{10}$

10. $2\frac{5}{6} \div 1\frac{1}{6} = \frac{17}{6} \div \frac{7}{6} =$
$\frac{17}{6} \times \frac{6}{7} = \frac{17}{7} = 2\frac{3}{7}$

11. $\frac{5}{8} \div \frac{1}{4} = \frac{5}{8} \times \frac{4}{1} = \frac{5}{2} = 2\frac{1}{2}$

12. $6\ m + 2\ m + 6\ m + 2\ m = 16\ m$

13. $1.2' + 1.2' + 1.2' + 1.2' = 4.8'$

14. $17" + 10" + 17" + 10" = 54"$

15. $\$10.95 \times 1.20 = \$13.14$

16. $3.14(15\ in)^2 = 706.5\ in^2$

17. $\$20.50 + \$15.98 + \$11.24 = \$47.72$
$\$47.72 \times 1.06 = \$50.58$

18. $2\frac{1}{2}$ cups $\times 1\frac{1}{2} = \frac{5}{2}$ cups $\times \frac{3}{2}$
    $= \frac{15}{4}$ cups $= 3\frac{3}{4}$ cups

19. $8\frac{1}{4} \div \frac{3}{4} = \frac{\cancel{33}^{11}}{\cancel{4}} \times \frac{\cancel{4}}{\cancel{3}} = 11$ gifts

20. 100 m = 100,000 mm = 0.1 km

Lesson Practice 17A

1. done
2. done
3.
    ```
       0.587
    6 ) 3.522
        3 0
          52
          48
          42
          42
    ```
4. $6 \times 0.587 = 3.522$
5.
    ```
       8.8
    9 ) 79.2
        72
        72
        72
    ```
6. $9 \times 8.8 = 79.2$
7.
    ```
       0.8
    5 ) 4.0
        4 0
    ```
8. $5 \times 0.8 = 4$
9.
    ```
       0.004
    2 ) 0.008
           8
    ```
10. $2 \times 0.004 = 0.008$
11.
    ```
       0.0125
    4 ) 0.0500
         4
         10
          8
          20
    ```
12. $4 \times 0.0125 = 0.05$
13. $3.6 \div 4 = 0.9$ pies
14. 5 pints $\div 8 = 0.625$ pints

15. $\$0.75 \div 5 = \$0.15$

Lesson Practice 17B

1.
    ```
       1.5
    4 ) 6.0
        4
        2 0
        2 0
    ```
2. $4 \times 1.5 = 6$
3.
    ```
       0.09
    3 ) 0.27
         27
    ```
4. $3 \times 0.09 = 0.27$
5.
    ```
       0.006
    7 ) 0.042
          42
    ```
6. $7 \times 0.006 = 0.042$
7.
    ```
       0.05
    8 ) 0.40
         40
    ```
8. $8 \times 0.05 = 0.4$
9.
    ```
       0.203
    6 ) 1.218
        1 2
        018
         18
    ```
10. $6 \times 0.203 = 1.218$
11.
    ```
       42.8
    2 ) 85.6
        8
        5
        4
        1 6
        1 6
    ```
12. $2 \times 42.8 = 85.6$
13. $24.8 \div 2 = 12.4$ pieces;
    12 pieces with 0.4 of a
    piece left over
14. 2.4 bushels $\div 3 = 0.8$ bushels
15. 34.4 gal $\times 4$ qt/gal $= 137.6$ qt; 137 jars

Lesson Practice 17C

1. $5 \overline{)0.40}$ = 0.08; 40
2. $5 \times 0.08 = 0.4$
3. $6 \overline{)0.408}$ = 0.068; 36, 48, 48
4. $6 \times 0.068 = 0.408$
5. $8 \overline{)0.064}$ = 0.008; 64
6. $8 \times 0.008 = 0.064$
7. $4 \overline{)1.2}$ = 0.3; 12
8. $4 \times 0.3 = 1.2$
9. $9 \overline{)28.08}$ = 3.12; 27, 10, 9, 18, 18
10. $9 \times 3.12 = 28.08$
11. $6 \overline{)3.0}$ = 0.5; 30
12. $6 \times 0.5 = 3$
13. 345.5 miles ÷ 2 days = $\dfrac{345.5 \text{ mi}}{2 \text{ days}}$
    172.75 miles per day (mi/day)
14. 125.1 bushels ÷ 9 days = 13.9 bushels/day
15. $24.90 ÷ 3 = $8.30 per child

Systematic Review 17D

1. $7 \overline{)0.14}$ = 0.02; 14
2. $7 \times 0.02 = 0.14$
3. $3 \overline{)3.6}$ = 1.2; 3, 6, 6
4. $3 \times 1.2 = 3.6$
5. $9 \overline{)0.549}$ = 0.061; 54, 9, 9
6. $9 \times 0.061 = 0.549$
7. $8 \overline{)5.000}$ = 0.625; 4 8, 20, 16, 40, 40
8. $8 \times 0.625 = 5$
9. $3.14(2.5 \text{ mi})^2 = 19.625 \text{ mi}^2$
10. $3.14(5 \text{ mi}) = 15.7 \text{ mi}$
11. 0.015 hg
12. 0.005 kl
13. 160 cm
14. done
15. 13' + 3' + 13' + 3' = 32'
16. 7.2 m + 4.1 m + 7.2 m + 4.1 m = 22.6 m
17. $32.50 ÷ 2 = $16.25
18. $2\dfrac{3}{4} - \dfrac{5}{8} = 2\dfrac{6}{8} - \dfrac{5}{8} = 2\dfrac{1}{8}$ pizza

Systematic Review 17E

1. $4 \overline{)0.032}$ = 0.008; 32
2. $4 \times 0.008 = 0.032$
3. $6 \overline{)0.30}$ = 0.05; 30
4. $6 \times 0.05 = 0.3$

Systematic Review 17E

5. $\quad\begin{array}{r}11.1\\5\overline{)55.5}\\\underline{5}\\5\\\underline{5}\\5\\\underline{5}\end{array}$

6. $5 \times 11.1 = 55.5$

7. $\quad\begin{array}{r}0.515\\2\overline{)1.030}\\\underline{1\ 0}\\3\\\underline{2}\\10\\\underline{10}\end{array}$

8. $2 \times 0.515 = 1.030$
9. $3.14(8\text{ km})^2 = 200.96\text{ km}^2$
10. $2(3.14)(8\text{ km}) = 50.24\text{ km}$
11. 1 g
12. 0.003 km
13. 2.5 dl
14. $4m + 3.8m + 4m + 3.8m = 15.6\text{ m}$
15. $25" + 9.1" + 25" + 9.1" = 68.2"$
16. $0.12\text{ km} + 0.05\text{ km} + 0.12\text{ km} + 0.05\text{ km} = 0.34\text{ km}$
17. $\$15.00 \times 0.9 = \$13.50$ sale price
18. $\$13.50 \times 1.04 = \$14.04$ total cost
    $\$15.00 - \$14.04 = \$0.96$ left over
19. $5.7\text{ km} \div 3 = 1.9\text{ km}$
20. $10\text{ lb} - 2\frac{5}{8}\text{ lb} = 9\frac{8}{8}\text{ lb} - 2\frac{5}{8}\text{ lb} = 7\frac{3}{8}\text{ lb}$

Systematic Review 17F

1. $\quad\begin{array}{r}0.001\\1\overline{)0.001}\\\underline{1}\end{array}$

2. $1 \times 0.001 = 0.001$

3. $\quad\begin{array}{r}0.32\\7\overline{)2.24}\\\underline{2\ 1}\\14\\\underline{14}\end{array}$

4. $7 \times 0.32 = 2.24$

5. $\quad\begin{array}{r}0.05\\3\overline{)0.15}\\\underline{15}\end{array}$

6. $3 \times 0.05 = 0.15$

7. $\quad\begin{array}{r}0.125\\8\overline{)1.000}\\\underline{8}\\20\\\underline{16}\\40\\\underline{40}\end{array}$

8. $8 \times 0.125 = 1$
9. $3.14(2\text{ mi})^2 = 12.56\text{ mi}^2$
10. $2(3.14)(2\text{ mi}) = 12.56\text{ mi}$
11. 0.95 hl
12. 2 cm
13. 40 cg
14. $80" + 70" + 80" + 70" = 300"$
15. $3.1m + 1.3m + 3.1m + 1.3m = 8.8\text{ m}$
16. $5\frac{1}{2}" + 3\frac{1}{2}" + 5\frac{1}{2}" + 3\frac{1}{2}" = 18"$
17. $5{,}000\text{ m} = 5\text{ km} = 500{,}000\text{ cm}$
18. $\$5.76 - \$0.89 = \$4.87$
19. $\$38.40 \times 1.15 = \$44.16$ total
    $\$44.16 \div 4 = \$11.04$ each
20. $5\frac{1}{2} - 2\frac{2}{3} = 5\frac{3}{6} - 2\frac{4}{6} =$
    $4\frac{9}{6} - 2\frac{4}{6} = 2\frac{5}{6}$ rows

Lesson Practice 18A

1. done
2. done

3. $\quad\begin{array}{r}300\\6\overline{)1800}\\\underline{1800}\end{array}$

4. $0.06 \times 300 = 18$

5. $\quad\begin{array}{r}1{,}000\\5\overline{)5{,}000}\\\underline{5{,}000}\end{array}$

6. $0.5 \times 1{,}000 = 500$

LESSON PRACTICE 18A - LESSON PRACTICE 18C

7. $9{,}000$
    $7\,\overline{)63{,}000}$
    $\underline{63{,}000}$

8. $0.007 \times 9{,}000 = 63$

9. 120
    $9\,\overline{)1{,}080}$
    $\underline{9}$
    180
    $\underline{180}$

10. $0.9 \times 120 = 108$

11. $1{,}600$
    $25\,\overline{)40{,}000}$
    $\underline{25}$
    $15{,}000$
    $\underline{15{,}000}$

12. $0.25 \times 1{,}600 = 400$

13. 80
    $5\,\overline{)400}$
    $\underline{400}$

    40 people ÷ 0.5 lb/person = 80 people

14. 80
    $3\,\overline{)240}$
    $\underline{240}$

    24 rows ÷ 0.3 rows/day = 80 days

15. 30
    $8\,\overline{)240}$
    $\underline{240}$

    24 rows ÷ 0.8 rows/day = 30 days

Lesson Practice 18B

1. $6{,}320$
    $1\,\overline{)6{,}320}$
    $\underline{6{,}320}$

2. $0.1 \times 6{,}320 = 632$

3. $2{,}200$
    $15\,\overline{)33{,}000}$
    $\underline{30}$
    $3{,}000$
    $\underline{3{,}000}$

4. $0.15 \times 2{,}200 = 330$

5. 230
    $6\,\overline{)1{,}380}$
    $\underline{1\,2}$
    180
    $\underline{180}$

6. $0.6 \times 230 = 138$

7. $2{,}000$
    $33\,\overline{)66{,}000}$
    $\underline{66{,}000}$

8. $0.033 \times 2{,}000 = 66$

9. 530
    $4\,\overline{)2{,}120}$
    $\underline{2{,}000}$
    120
    $\underline{120}$

    (Filling in extra zeros to mark place value is optional when working a division problem.)

10. $0.4 \times 530 = 212$

11. 500
    $75\,\overline{)37{,}500}$
    $\underline{37{,}500}$

12. $0.75 \times 500 = 375$

13. $\$200 \div \$0.50 = 400$ gifts

14. 36 yd ÷ 0.9 yd/pillow = 40 pillows

15. 3 tons ÷ 0.01 tons/day = 300 days

Lesson Practice 18C

1. $2{,}290$
    $2\,\overline{)4{,}580}$
    $\underline{4{,}000}$
    580
    $\underline{400}$
    180
    $\underline{180}$

2. $0.2 \times 2{,}290 = 458$

3. 200
    $8\,\overline{)1{,}600}$
    $\underline{1{,}600}$

4. $0.08 \times 200 = 16$

5.
```
      160
   7 )1,120
      700
      420
      420
```

6. $0.7 \times 160 = 112$

7.
```
       38,000
   9 )342,000
      270,000
       72,000
       72,000
```

8. $0.009 \times 38,000 = 342$

9.
```
       1,190
   3 )3,570
      3,000
        570
        300
        270
        270
```

10. $0.3 \times 1,190 = 357$

11.
```
       1,000
   45 )45,000
       45,000
```

12. $0.45 \times 1,000 = 450$
13. $5\ L \div 0.05\ L/batch = 100$ batches
14. $\$35.00 \div \$0.25/quarter = 140$ quarters
15. $244\ lb \div 0.4\ lb/cust. = 560$ customers

Systematic Review 18D

1.
```
      60
   6 )360
      360
```

2. $0.6 \times 60 = 36$

3.
```
       2,000
   9 )18,000
      18,000
```

4. $0.09 \times 2,000 = 180$

5.
```
       3.54
   2 )7.08
      6 00
        10
        10
         8
         8
```

6. $2 \times 3.54 = 7.08$

7.
```
        0.002
   15 )0.030
         30
```

8. $15 \times 0.002 = 0.03$
9. $3.14(5.5\ ft)^2 = 94.985\ ft^2$
10. $3.14(11\ ft) = 34.54\ ft$
11. $8 \times 8 = 64$
12. $10 \times 10 = 100$
13. $2 \times 2 = 4$
14. done
15. $3' + 4' + 5' = 12'$
16. $5.4' + 3.9' + 7.8' = 17.1'$
17. $14" + 14" + 14" + 14" = 56"$
18. $5\ tons \times 1.50 = 7.5\ tons$

Systematic Review 18E

1.
```
       1,000
   1 )1,000
      1,000
```

2. $0.1 \times 1,000 = 100$

3.
```
       250
   6 )1,500
      1 2
        300
        300
```

4. $0.06 \times 250 = 15$

5.
```
       1.9
   8 )15.2
      8 0
      7 2
      7 2
```

6. $8 \times 1.9 = 15.2$

SYSTEMATIC REVIEW 18E - LESSON PRACTICE 19A

7. $\phantom{30\,\overline{|}}0.0002$
   $30\,\overline{|\,0.0060}$
   $\phantom{30\,\overline{|\,0.00}}\underline{60}$

8. $30 \times 0.0002 = 0.006$

9. $3.14(9\text{ m})^2 = 254.34\text{ m}^2$

10. $2(3.14)(9\text{ m}) = 56.52\text{ m}$

11. $12 \times 12 = 144$

12. $1 \times 1 = 1$

13. $5 \times 5 = 25$

14. $8' + 8' + 10' = 26'$

15. $5'' + 12'' + 13'' = 30''$

16. $1.7\text{ m} + 1.1\text{ m} + 1.9\text{ m} = 4.7\text{ m}$

17. $20\text{ cm} + 30\text{ cm} + 20\text{ cm} + 30\text{ cm} = 100\text{ cm} = 1\text{ m}$

18. $100 \times 2.1 = 210$ nature books

19. $\frac{1}{2}c + 1\frac{1}{4}c = \frac{2}{4}c + 1\frac{1}{4}c = 1\frac{3}{4}$ cups

20. $\$8.00 \div \$0.75 = 10.67 = 10$ cones

Systematic Review 18F

1. $\phantom{3\,\overline{|}}123{,}000$
   $3\,\overline{|\,369{,}000}$
   $\phantom{3\,\overline{|}}\underline{300{,}000}$
   $\phantom{3\,\overline{|}\,3}69{,}000$
   $\phantom{3\,\overline{|}\,3}\underline{60{,}000}$
   $\phantom{3\,\overline{|}\,33}9{,}000$
   $\phantom{3\,\overline{|}\,33}\underline{9{,}000}$

2. $0.003 \times 123{,}000 = 369$

3. $\phantom{49\,\overline{|}}100$
   $49\,\overline{|\,4{,}900}$
   $\phantom{49\,\overline{|}}\underline{4{,}900}$

4. $0.49 \times 100 = 49$

5. $\phantom{5\,\overline{|}}4.5$
   $5\,\overline{|\,22.5}$
   $\phantom{5\,\overline{|}}\underline{200}$
   $\phantom{5\,\overline{|}\,2}25$
   $\phantom{5\,\overline{|}\,2}\underline{25}$

6. $5 \times 4.5 = 22.5$

7. $\phantom{8\,\overline{|}}0.0002$
   $8\,\overline{|\,0.0016}$
   $\phantom{8\,\overline{|}}\underline{16}$

8. $8 \times 0.0002 = 0.0016$

9. $3.14(1.2\text{ in})^2 = 4.5216\text{ in}^2$

10. $2(3.14)(1.2\text{ in}) = 7.536\text{ in}$

11. $25 \times 25 = 625$

12. $7 \times 7 = 49$

13. $100 \times 100 = 10{,}000$

14. $25' + 25' + 32' = 82'$

15. $6'' + 8'' + 10'' = 24''$

16. $0.89\text{ m} + 0.7\text{ m} + 1.08\text{ m} = 2.67\text{ m}$

17. $25\text{ lb} \div 0.5\text{ lb/package} = 50$ packages

18. $\$37.50 \times 1.50 = \$56.25$
    $\$56.25 \div 50$ package $= \$1.125$/package;
    rounds to $\$1.13$ per package

19. $\$42.50 \times 1.34 = \$56.95$

20. $4{,}520\text{ g} = 4.52\text{ kg}$

Lesson Practice 19A

1. done

2. done

3. $0.06X = 24$
   $\dfrac{0.06X}{0.06} = \dfrac{24}{0.06}$
   $X = 24 \div 0.06 = 400$

4. $0.06(400) = 24$
   $24 = 24$

5. $0.7A = 490$
   $\dfrac{0.7A}{0.7} = \dfrac{490}{0.7}$
   $A = 490 \div 0.7 = 700$

6. $0.7(700) = 490$
   $490 = 490$

7. $0.002R = 6$
   $\dfrac{0.002R}{0.002} = \dfrac{6}{0.002}$
   $R = 6 \div 0.002 = 3{,}000$

8. $0.002(3{,}000) = 6$
   $6 = 6$

9. $0.2M = \$84$
   $\dfrac{0.2M}{0.2} = \dfrac{\$84}{0.2}$
   $M = \$84 \div 0.2 = \$420.00$

LESSON PRACTICE 19A - SYSTEMATIC REVIEW 19D

10. $0.15T = 3'$

    $\dfrac{0.15T}{0.15} = \dfrac{3'}{0.15}$

    $T = 3' \div 0.15 = 20'$

Lesson Practice 19B

1. $0.5Y = 8$

   $\dfrac{0.5Y}{0.5} = \dfrac{8}{0.5}$

   $Y = 8 \div 0.5 = 16$

2. $0.5(16) = 8$
   $8 = 8$

3. $0.03X = 21$

   $\dfrac{0.03X}{0.03} = \dfrac{21}{0.03}$

   $X = 21 \div 0.03 = 700$

4. $0.03(700) = 21$
   $21 = 21$

5. $0.1A = 30$

   $\dfrac{0.1A}{0.1} = \dfrac{30}{0.1}$

   $A = 30 \div 0.1 = 300$

6. $0.1(300) = 30$
   $30 = 30$

7. $0.008R = 16$

   $\dfrac{0.008R}{0.008} = \dfrac{16}{0.008}$

   $R = 16 \div 0.008 = 2{,}000$

8. $0.008(2{,}000) = 16$
   $16 = 16$

9. $0.7P = \$140$

   $\dfrac{0.7P}{0.7} = \dfrac{\$140}{0.7}$

   $P = \$140 \div 0.7 = \$200.00$

10. $0.8H = 60"$

    $\dfrac{0.8H}{0.8} = \dfrac{60"}{0.8}$

    $H = 60" \div 0.8 = 75"$

Lesson Practice 19C

1. $0.6Y = 6$

   $\dfrac{0.6Y}{0.6} = \dfrac{6}{0.6}$

   $Y = 6 \div 0.6 = 10$

2. $0.6(10) = 6$
   $6 = 6$

3. $0.04X = 32$

   $\dfrac{0.04X}{0.04} = \dfrac{32}{0.04}$

   $X = 32 \div 0.04 = 800$

4. $0.04(800) = 32$
   $32 = 32$

5. $0.3A = 93$

   $\dfrac{0.3A}{0.3} = \dfrac{93}{0.3}$

   $A = 93 \div 0.3 = 310$

6. $0.3(310) = 93$
   $93 = 93$

7. $0.005R = 5$

   $\dfrac{0.005R}{0.005} = \dfrac{5}{0.005}$

   $R = 5 \div 0.005 = 1{,}000$

8. $0.005(1{,}000) = 5$
   $5 = 5$

9. $0.25G = 4 \text{ bushels}$

   $\dfrac{0.25G}{0.25} = \dfrac{4 \text{ bushels}}{0.25}$

   $G = 4 \text{ bushels} \div 0.25$
   $G = 16 \text{ bushels}$

10. $0.05R = \$10$

    $\dfrac{0.05R}{0.05R} = \dfrac{\$10}{0.05}$

    $R = \$10 \div 0.05 = \$200.00$

Systematic Review 19D

1. $0.22Q = 88$

   $\dfrac{0.22Q}{0.22} = \dfrac{88}{0.22}$

   $Q = 88 \div 0.22 = 400$

SYSTEMATIC REVIEW 19D - SYSTEMATIC REVIEW 19F

2. $0.22(400) = 88$
   $88 = 88$
3. $0.007X = 35$
   $\dfrac{0.007X}{0.007} = \dfrac{35}{0.007}$
   $X = 35 \div 0.007 = 5{,}000$
4. $0.007(5{,}000) = 35$
   $35 = 35$
5. $4 \div 20 = 0.2$
6. $20 \times 0.2 = 4$
7. $3.6 \div 18 = 0.2$
8. $18 \times 0.2 = 3.6$
9. done
10. $10 \text{ ft} \times 10 \text{ ft} = 100 \text{ ft}^2$
11. $15 \text{ m} \times 7 \text{ m} = 105 \text{ m}^2$
12. $9 \text{ ft} \times 12 \text{ ft} = 108 \text{ ft}^2$
13. $9' + 12' + 9' + 12' = 42'$
14. $0.35M = \$105$
    $\dfrac{0.35M}{0.35} = \dfrac{\$105}{0.35}$
    $M = \$105 \div 0.35 = \$300.00$
15. $2(3.14)(5 \text{ mi}) = 31.4 \text{ mi}$
16. $3.14(5 \text{ mi})^2 = 78.5 \text{ mi}^2$
17. $5 \text{ km} = 5{,}000 \text{ m}$
    $5{,}000 \text{ m} > 500 \text{ m}$; she drove farther.
18. $\$30.75 \div 5 = \$6.15$

Systematic Review 19E

1. $0.9D = 27$
   $\dfrac{0.9D}{0.9} = \dfrac{27}{0.9}$
   $D = 27 \div 0.9 = 30$
2. $0.9(30) = 27$
   $27 = 27$
3. $0.08F = 5$
   $\dfrac{0.08F}{0.08} = \dfrac{5}{0.08}$
   $F = 5 \div 0.08 = 62.5$
4. $0.08(62.5) = 5$
   $5 = 5$

5. $5{,}500 \div 11 = 500$
6. $0.11 \times 500 = 55$
7. $0.16 \div 40 = 0.004$
8. $40 \times 0.004 = 0.16$
9. $10 \text{ in} \times 4 \text{ in} = 40 \text{ in}^2$
10. $45 \text{ ft} \times 45 \text{ ft} = 2{,}025 \text{ ft}^2$
11. $2.3 \text{ m} \times 1.2 \text{ m} = 2.76 \text{ m}^2$
12. $4 \text{ yd} \times 11 \text{ yd} = 44 \text{ yd}^2$
13. $3 \text{ ft} + 3 \text{ ft} + 3 \text{ ft} + 3 \text{ ft} = 12 \text{ ft}$
    $12 \times 1 = 12$ flowers
14. $0.40P = \$80$
    $\dfrac{0.40P}{0.40} = \dfrac{\$80}{0.40}$
    $P = \$80 \div 0.40 = \$200.00$
15. $10 \text{ in} \times 12 \text{ in} = 120 \text{ in}^2$
16. $3.14(6 \text{ in})^2 = 113.04 \text{ in}^2$
17. $120 \text{ in}^2 > 113.04 \text{ in}^2$
    The rectangular one would be better.
18. $2 \text{ m} = 200 \text{ cm}$
    They are the same height.
19. $\$57.04 \div 8 = \$7.13$
20. $4.5" + 5" + 3.75" = 13.25"$

Systematic Review 19F

1. $0.32G = 64$
   $\dfrac{0.32G}{0.32} = \dfrac{64}{0.32}$
   $G = 64 \div 0.32 = 200$
2. $0.32(200) = 64$
   $64 = 64$
3. $0.005Y = 4$
   $\dfrac{0.005Y}{0.005} = \dfrac{4}{0.005}$
   $Y = 4 \div 0.005 = 800$
4. $0.005(800) = 4$
   $4 = 4$
5. $10{,}000 \div 1 = 10{,}000$
6. $0.001 \times 10{,}000 = 10$
7. $1.26 \div 21 = 0.06$
8. $0.06 \times 21 = 1.26$

9. $6 \text{ m} \times 2 \text{ m} = 12 \text{ m}^2$
10. $1.2 \text{ ft} \times 1.2 \text{ ft} = 1.44 \text{ ft}^2$
11. $17 \text{ in} \times 10 \text{ in} = 170 \text{ in}^2$
12. $5 \text{ ft} \times 6 \text{ ft} = 30 \text{ ft}^2$
13. $3.14(3 \text{ ft})^2 = 28.26 \text{ ft}^2$
14. $30 \text{ ft}^2 - 28.26 \text{ ft}^2 = 1.74 \text{ ft}^2$
15. $1 \text{ mi} + 1 \text{ mi} + 1 \text{ mi} + 1 \text{ mi} = 4 \text{ mi}$
16. $0.32T = 200 \text{ hr}$
    $$\frac{0.32T}{0.32} = \frac{200 \text{ hr}}{0.32}$$
    $T = 200 \text{ hr} \div 0.32 = 625 \text{ hr}$
17. $625 \text{ hr} - 200 \text{ hr} = 425 \text{ hr}$
18. $652 \text{ g} = 0.652 \text{ kg}$
19. $\$121.92 \div 12 = \$10.16$
20. $6.25" + 7.1" + 6.25" + 7.1" = 26.7"$

Lesson Practice 20A

1. done
2. done
3. done
4. done
5.
    ```
        9
    9 )81
       81
    ```
6. $0.9 \times 9 = 8.1$
7.
    ```
       40
    6 )240
      240
    ```
8. $0.006 \times 40 = 0.24$
9.
    ```
       59
    8 )472
      472
    ```
10. $0.8 \times 59 = 47.2$
11.
    ```
          2.72
    15 )40.80
        30 00
        10 80
        10 50
            30
            30
    ```

12. $1.5 \times 2.72 = 4.080$
13. $1.8 \text{ L} \div 0.3 \text{ L/guest} = 6 \text{ guests}$
14. $2.6 \text{ acres} \div 1.3 \text{ acres/day} = 2 \text{ days}$
15. $6.9 \text{ pt} \div 2.3 \text{ pt/person} = 3 \text{ people}$

Lesson Practice 20B

1.
    ```
        0.09
    5 )0.45
        45
    ```
2. $0.5 \times 0.09 = 0.045$
3.
    ```
          5
    82 )410
        410
    ```
4. $0.82 \times 5 = 4.1$
5.
    ```
        1.1
    2 )2.2
      2 0
        2
        2
    ```
6. $0.2 \times 1.1 = 0.22$
7.
    ```
        71
    9 )639
      630
        9
        9
    ```
8. $0.009 \times 71 = 0.639$
9.
    ```
        61.2
    7 )428.4
      420 0
        8 4
        7 0
        1 4
        1 4
    ```
10. $0.7 \times 61.2 = 42.84$
11.
    ```
         15
    21 )315
        210
        105
        105
    ```
12. $2.1 \times 15 = 31.5$
13. $0.856 \text{ kg} \div 0.4 \text{ kg/sack} = 2.14 \text{ sacks}$
    He has 3 sacks.

LESSON PRACTICE 20B - SYSTEMATIC REVIEW 20D

14. 5.5 lb ÷ 0.25 lb/favor = 22 favors
15. $45.50 ÷ $0.35 per item = 130 items

Lesson Practice 20C

1. 0.06
    $8\,\overline{)0.48}$
    $\underline{48}$

2. $0.8 \times 0.06 = 0.048$

3. 40.6
    $22\,\overline{)893.2}$
    $\underline{8800}$
    $13\,2$
    $\underline{13\,2}$

4. $0.22 \times 40.6 = 8.932$

5. 13
    $5\,\overline{)65}$
    $\underline{50}$
    15
    $\underline{15}$

6. $0.5 \times 13 = 6.5$

7. 580
    $7\,\overline{)4060}$
    $\underline{3500}$
    560
    $\underline{560}$

8. $0.007 \times 580 = 4.06$

9. 91.2
    $4\,\overline{)364.8}$
    $\underline{360\,0}$
    $4\,8$
    $\underline{4\,8}$

10. $0.4 \times 91.2 = 36.48$

11. 16
    $32\,\overline{)512}$
    $\underline{320}$
    192
    $\underline{192}$

12. $3.2 \times 16 = 51.2$
13. 25.5 lb ÷ 0.5 lb/bag = 51 bags
14. 4.6 mi ÷ 0.2 mi/rest = 23 rests
15. $16.80 ÷ $2.10 / item = 8 items

Systematic Review 20D

1. 0.5
    $3\,\overline{)1.5}$
    $\underline{1\,5}$

2. $0.3 \times 0.5 = 0.15$

3. 1.6
    $25\,\overline{)40.0}$
    $\underline{25\,0}$
    $15\,0$
    $\underline{15\,0}$

4. $0.025 \times 1.6 = 0.04$

5. 0.002
    $60\,\overline{)0.120}$
    $\underline{120}$

6. $60 \times 0.002 = 0.12$

7. 430
    $1\,\overline{)430}$
    $\underline{400}$
    30
    $\underline{30}$

8. $0.1 \times 430 = 43$

9. $0.8G = 40$
   $G = 40 \div 0.8 = 50$

10. $0.15Y = 0.3$
    $Y = 0.3 \div 0.15 = 2$

11. $8\frac{1}{2} + 2\frac{7}{8} = 8\frac{4}{8} + 2\frac{7}{8} = 10\frac{11}{8} = 11\frac{3}{8}$

12. $16 - 1\frac{2}{3} = 15\frac{3}{3} - 1\frac{2}{3} = 14\frac{1}{3}$

13. $3\frac{1}{8} \times 1\frac{1}{3} = \frac{25}{{}_2 8} \times \frac{\cancel{4}}{3} = \frac{25}{6} = 4\frac{1}{6}$

14. done
15. 13 in × 28 in = 364 in²
16. 2 m × 1.2 m = 2.4 m²
17. 220m + 150m + 220m + 150m = 740 m
    220m × 100m = 22,000 m²
18. $0.25M = 37.50
    $M = $37.50 \div 0.25 = 150.00

Systematic Review 20E

1. $$\begin{array}{r} 228 \\ 2\overline{)456} \\ \underline{400} \\ 56 \\ \underline{40} \\ 16 \\ \underline{16} \end{array}$$

2. $0.02 \times 228 = 4.56$

3. $$\begin{array}{r} 910 \\ 9\overline{)8{,}190} \\ \underline{8{,}100} \\ 90 \\ \underline{90} \end{array}$$

4. $0.009 \times 910 = 8.19$

5. $$\begin{array}{r} 5.01 \\ 12\overline{)60.12} \\ \underline{60\ 00} \\ 12 \\ \underline{12} \end{array}$$

6. $12 \times 5.01 = 60.12$

7. $$\begin{array}{r} 20 \\ 45\overline{)900} \\ \underline{900} \end{array}$$

8. $4.5 \times 20 = 90$

9. $6.2X = 7.44$
   $X = 7.44 \div 6.2 = 1.2$

10. $0.4B = 0.88$
    $B = 0.88 \div 0.4 = 2.2$

11. $6\frac{3}{8} \div 3\frac{1}{4} =$
    $\frac{51}{8} \div \frac{13}{4} =$
    $\frac{51}{{}_2 8} \times \frac{\cancel{4}}{13} = \frac{51}{26} = 1\frac{25}{26}$

12. $65 - 21\frac{5}{6} = 64\frac{6}{6} - 21\frac{5}{6} = 43\frac{1}{6}$

13. $9\frac{3}{10} \times 5\frac{2}{3} =$
    $\frac{\cancel{93}^{31}}{10} \times \frac{17}{\cancel{3}} = \frac{527}{10} = 52\frac{7}{10}$

14. $3.8 \text{ m} \times 3 \text{ m} = 11.4 \text{ m}^2$

15. $20 \text{ mm} \times 7 \text{ mm} = 140 \text{ mm}^2$

16. $25 \text{ in} \times 8.5 \text{ in} = 212.5 \text{ in}^2$

17. $25 \text{ yd} \times 33 \text{ yd} = 825 \text{ yd}^2 \text{ total}$
    $825 \text{ yd}^2 \times \frac{1}{3} = 275 \text{ yd}^2 \text{ planted}$

18. $275 \text{ yd}^2 \div 0.5 \text{ yd}^2/\text{plant} = 550 \text{ plants}$

19. $550 \text{ plants} \times \$0.85/\text{plant} = \$467.50$

20. $1.05A = \$67.41$
    $A = \$67.41 \div 1.05 = \$64.20$

Systematic Review 20F

1. $$\begin{array}{r} 78 \\ 4\overline{)312} \\ \underline{280} \\ 32 \\ \underline{32} \end{array}$$

2. $0.4 \times 78 = 31.2$

3. $$\begin{array}{r} 15 \\ 28\overline{)420} \\ \underline{280} \\ 140 \\ \underline{140} \end{array}$$

4. $0.28 \times 15 = 4.2$

5. $$\begin{array}{r} 0.056 \\ 35\overline{)1.960} \\ \underline{1\ 750} \\ 210 \\ \underline{210} \end{array}$$

6. $35 \times 0.056 = 1.96$

7. $$\begin{array}{r} 20 \\ 88\overline{)1760} \\ \underline{1760} \end{array}$$

8. $8.8 \times 20 = 176$

9. $0.11Q = 0.55$
   $Q = 0.55 \div 0.11 = 5$

10. $1.3W = 0.39$
    $W = 0.39 \div 1.3 = 0.3$

11. $\frac{6}{5} \div \frac{2}{5} = \frac{6 \div 2}{1} = 3$

12. $18 + 6\frac{7}{8} = 24\frac{7}{8}$

13. $\frac{11}{{}_3\cancel{12}} \times \frac{\cancel{4}}{5} = \frac{11}{15}$

SYSTEMATIC REVIEW 20F - LESSON PRACTICE 21A

14. 80 in × 55 in = 4,400 in²

15. 3.1 m × 0.9 m = 2.79 m²

16. $\frac{11}{2}$ in × $\frac{21}{8}$ in = $\frac{231}{16}$ in² = 14$\frac{7}{16}$ in²

17. $\pi(6\text{ ft})^2 \approx 113.04$ ft²

18. 5 k̶m̶ × $\frac{1,000 \text{ m}}{1 \text{ k̶m̶}}$ = 5,000 m

19. 1.15M = $18.86
    M = $18.86 ÷ 1.15 = $16.40

20. 9 mi + 12 mi + 15 mi = 36 mi

Lesson Practice 21A

1. done

2.
```
       0.712
   5 )3.560
       3 500
         60
         50
         10
         10
```

3.
```
        11.25
    8 )90.00
       80 00
       10 00
        8 00
        2 00
        1 60
          40
          40
```

4. done

5.
```
        1.973 ≈ 1.97
    3 )5.920
       3 000
       2 920
       2 700
         220
         210
          10
```

(≈ means "approximately equal to")

6.
```
         17.555 ≈ 17.56
   9 )158.000
       90 000
       68 000
       63 000
        5 000
        4 500
          500
          450
           50
           45
            5
```

7. done

8.
```
       7.33 or 7.3̄
   3 )22.00
      21 00
       1 00
         90
         10
          9
```

9.
```
        9.166 or 9.16̄
   6 )55.000
      54 000
       1 000
         600
         400
         360
          40
          36
```

10. done

11. 0.59$\frac{6}{8}$ = 0.59$\frac{3}{4}$
```
   8 )4.78
      4 00
        78
        72
         6
```

12.
```
       4.72  2/9
     _____
  9 | 42.50
     36 00
     ─────
      6 50
      6 30
      ────
        20
        18
        ──
         2
```

Lesson Practice 21B

1.
```
       0.2335
     _____
  2 | 0.4670
     4000
     ────
      670
      600
      ───
       70
       60
       ──
       10
       10
```

2.
```
       12.75
     _____
  4 | 51.00
     40 00
     ─────
     11 00
      8 00
     ─────
      3 00
      2 80
      ────
        20
        20
```

3.
```
        15.4
     _____
  5. | 77.0
      5 00
      ────
      2 70
      2 50
      ────
        20
        20
```

4.
```
        4.557 ≈ 4.56
     _____
  7 | 31.900
     28 000
     ──────
      3 900
      3 500
      ─────
        400
        350
        ───
         50
         49
```

5.
```
        1.093 ≈ 1.09
     _____
  9 | 9.840
     9 000
     ─────
       840
       810
       ───
        30
        27
```

6.
```
        50.833 ≈ 50.83
     _____
  6 | 305.000
     300 000
     ───────
       5 000
       4 800
       ─────
         200
         180
         ───
          20
          18
```

7.
```
        0.133 or 0.1̄3
     _____
  3 | 0.400
     300
     ───
     100
      90
     ───
      10
       9
```

LESSON PRACTICE 21B - LESSON PRACTICE 21C

8. $\quad\quad\quad$ 0.01428571 or 0.0$\overline{142857}$
    7) 0.10000000
    $\quad\quad\quad\underline{7000000}$
    $\quad\quad\quad$ 3000000
    $\quad\quad\quad\underline{2800000}$
    $\quad\quad\quad$ 200000
    $\quad\quad\quad\underline{140000}$
    $\quad\quad\quad$ 60000
    $\quad\quad\quad\underline{56000}$
    $\quad\quad\quad$ 4000
    $\quad\quad\quad\underline{3500}$
    $\quad\quad\quad$ 500
    $\quad\quad\quad\underline{490}$
    $\quad\quad\quad$ 10
    $\quad\quad\quad\underline{7}$

9. $\quad\quad\quad$ 5.9166 or 5.91$\overline{6}$
    6) 35.5000
    $\quad\underline{30\ 0000}$
    $\quad$ 5 5000
    $\quad\underline{5\ 4000}$
    $\quad$ 1000
    $\quad\underline{600}$
    $\quad$ 400
    $\quad\underline{360}$
    $\quad$ 40
    $\quad\underline{36}$

10. $\quad\quad\quad$ 14.07 $\frac{1}{7}$
    7) 98.50
    $\underline{70\ 00}$
     28 50
    $\underline{28\ 00}$
     50
    $\underline{49}$
     1

11. $\quad\quad\quad$ 244.66 $\frac{2}{3}$
    3) 734.00
    $\underline{600\ 00}$
    134 00
    $\underline{120\ 00}$
     14 00
    $\underline{12\ 00}$
     2 00
    $\underline{1\ 80}$

12. $\quad\quad\quad$ 13.87 $\frac{4}{8}$ = 13.87 $\frac{1}{2}$
    8) 111.00
    $\underline{80\ 00}$
    31 00
    $\underline{24\ 00}$
     7 00
    $\underline{6\ 40}$
     60
    $\underline{56}$
     4

Lesson Practice 21C

1. $\quad\quad\quad$ 7.1375
    8) 57.1000
    $\underline{56\ 0000}$
     1 1000
    $\underline{8000}$
     3000
    $\underline{2400}$
     600
    $\underline{560}$
     40
    $\underline{40}$

2. $\quad\quad\quad$ 19.5
    2) 39.0
    $\underline{200}$
    190
    $\underline{180}$
     10
    $\underline{10}$

3. $\quad\quad\quad$ 3.125
    4) 12.500
    $\underline{12\ 000}$
     500
    $\underline{400}$
     100
    $\underline{80}$
     20
    $\underline{20}$

LESSON PRACTICE 21C - SYSTEMATIC REVIEW 21D

4. $2.266 \approx 2.27$
   $3\,\overline{)6.800}$
   $\underline{6\,000}$
   800
   $\underline{600}$
   200
   $\underline{180}$
   20
   $\underline{18}$

5. $0.742 \approx 0.74$
   $7\,\overline{)5.200}$
   $\underline{4\,900}$
   300
   $\underline{280}$
   20
   $\underline{14}$

6. $0.535 \approx 0.54$
   $9\,\overline{)4.820}$
   $\underline{4\,500}$
   320
   $\underline{270}$
   50
   $\underline{45}$

7. 0.4077 or $0.40\overline{7}$
   $9\,\overline{)3.6700}$
   $\underline{3\,6000}$
   700
   $\underline{630}$
   70
   $\underline{63}$

8. 0.181 or $0.1\overline{8}$
   $11\,\overline{)2.000}$
   $\underline{1\,100}$
   900
   $\underline{880}$
   20
   $\underline{11}$

9. 15.0166 or $15.01\overline{6}$
   $6\,\overline{)90.1000}$
   $\underline{60\,0000}$
   $30\,1000$
   $\underline{30\,0000}$
   1000
   $\underline{600}$
   400
   $\underline{360}$
   40
   $\underline{36}$

10. $0.78\frac{2}{10} = 0.78\frac{1}{5}$
    $10\,\overline{)7.82}$
    $\underline{7\,00}$
    82
    $\underline{80}$
    2

11. $0.04\frac{5}{7}$
    $7\,\overline{)0.33}$
    $\underline{28}$
    5

12. $2.04\frac{2}{12} = 2.04\frac{1}{6}$
    $12\,\overline{)24.50}$
    $\underline{24\,00}$
    50
    $\underline{48}$
    2

Systematic Review 21D

1. $0.178 \approx 0.18$
   $7\,\overline{)1.250}$
   $\underline{700}$
   550
   $\underline{490}$
   60
   $\underline{56}$

SYSTEMATIC REVIEW 21D - SYSTEMATIC REVIEW 21D

2. $1.666 \approx 1.67$

   ```
   3 ) 5.000
       3 000
       2 000
       1 800
         200
         180
          20
          18
   ```

3. $5.483 \approx 5.48$

   ```
   6 ) 32.900
       30 000
        2 900
        2 400
          500
          480
           20
           18
   ```

4. 0.9344 or $0.93\overline{4}$

   ```
   9 ) 8.4100
       8 1000
         3100
         2700
          400
          360
           40
           36
   ```

5. 06.33 or $6.\overline{3}$

   ```
   6 ) 38.00
       36 00
        2 00
        1 80
          20
          18
   ```

6. 0.2857142 or $0.\overline{285714}$

   ```
   7 ) 2.0000000
       1 4000000
         6000000
         5600000
          400000
          350000
           50000
           49000
            1000
             700
             300
             280
              20
              14
   ```

7. $8X = 0.5$

   $X = 0.5 \div 8 = 0.06\frac{2}{8} = 0.06\frac{1}{4}$

8. $0.3Y = 5.63$

   $Y = 5.63 \div 0.3 = 18.76\frac{2}{3}$

9. $10F = 0.46$

   $F = 0.46 \div 10 = 0.04\frac{6}{10} = 0.04\frac{3}{5}$

10. $0.05 + 1.9 = 1.95$

11. $3.67 - 0.08 = 3.59$

12. $0.006 + 4.2 = 4.206$

13. done

14. $\left(\frac{1}{2}\right)(9 \text{ in})(12 \text{ in}) = 54 \text{ in}^2$

15. $\left(\frac{1}{2}\right)(3.08 \text{ m})(0.7 \text{ m}) = 1.078 \text{ m}^2$

16. $\$355 \times 0.20 = \$71.00$

17. $2{,}400 \text{ g} \times \frac{1 \text{ kg}}{1{,}000 \text{ g}} = 2.4 \text{ kg}$

    $25 \text{ kg} - 2.4 \text{ kg} = 22.6 \text{ kg}$

18. $7H = 37.1 \text{ ft}^2$

    $H = 37.1 \text{ ft}^2 \div 7 \text{ ft} = 5.3 \text{ ft}$

Systematic Review 21E

1. $6.844 \approx 6.84$
$$9\overline{)61.600}$$
   54 000
   7 600
   7 200
   400
   360
   40
   36

2. $0.018 \approx 0.02$
$$20\overline{)0.370}$$
   200
   170
   160

3. $0.133 \approx 0.13$
$$3\overline{)0.400}$$
   300
   100
   90
   10
   9

4. 0.266 or $0.2\overline{6}$
$$6\overline{)1.600}$$
   1 200
   400
   360
   40
   36

5. 23.33 or $23.\overline{3}$
$$3\overline{)70.00}$$
   60 00
   10 00
   9 00
   1 00
   90

6. 0.10909 or $0.1\overline{09}$
$$11\overline{)1.20000}$$
   1 10000
   10000
   9900
   100
   99

7. $12X = 0.28$
$X = 0.28 \div 12 = 0.02\frac{4}{12} = 0.02\frac{1}{3}$

8. $0.9Y = 0.5$
$Y = 0.5 \div 0.9 = 0.55\frac{5}{9}$

9. $1.3F = 5.4$
$F = 5.4 \div 1.3 = 4.15\frac{5}{13}$

10. $6.15 - 0.06 = 6.09$
11. $14.003 + 0.5 = 14.503$
12. $8.3 - 0.67 = 7.63$
13. $\frac{1}{2}(4\,\text{ft})(6\,\text{ft}) = 12\,\text{ft}^2$
14. $\frac{1}{2}(12\,\text{in})(16\,\text{in}) = 96\,\text{in}^2$
15. $\frac{1}{2}(1.02\,\text{m})(0.5\,\text{m}) = 0.255\,\text{m}^2$
16. $\$11.00 \times 1.06 = \$11.66$
17. $\$0.96 \div 12\,\text{oz} = \$0.08$ per oz
18. $\$0.56 \div 8\,\text{oz} = \$0.07$ per oz
    The 8-ounce can is a better buy.
19. $4{,}180\,\text{m} \times \dfrac{1\,\text{km}}{1{,}000\,\text{m}} = 4.18\,\text{km}$
    $4.18\,\text{km} - 4\,\text{km} = 0.18\,\text{km}$
20. $20H = 40.22\,\text{in}^2$
    $H = 40.22\,\text{in}^2 \div 20\,\text{in}$
    $H = 2.011\,\text{in}$ (rounds to 2.01 in)

Systematic Review 21F

1. $8.062 \approx 8.06$
$$8\overline{)64.500}$$
   64 000
   500
   480
   20
   16

2. $0.304 \approx 0.30$
$$15\overline{)4.570}$$
   4 500
   70
   60

SYSTEMATIC REVIEW 21F - LESSON PRACTICE 22A

3. $\begin{array}{r} 2.128 \approx 2.13 \\ 7\overline{)14.900} \\ \underline{14\,000} \\ 900 \\ \underline{700} \\ 200 \\ \underline{140} \\ 60 \\ \underline{56} \\ \end{array}$

4. $\begin{array}{r} 2.242 \text{ or } 2.\overline{24} \\ 33\overline{)74.000} \\ \underline{66\,000} \\ 8\,000 \\ \underline{6\,600} \\ 1\,400 \\ \underline{1\,320} \\ 80 \\ \underline{66} \\ \end{array}$

5. $\begin{array}{r} 50.909 \text{ or } 50.\overline{90} \\ 11\overline{)560.000} \\ \underline{550\,000} \\ 10\,000 \\ \underline{9\,900} \\ 100 \\ \underline{99} \\ \end{array}$

6. $\begin{array}{r} 1.11 \text{ or } 1.\overline{1} \\ 9\overline{)10.00} \\ \underline{9\,00} \\ 1\,00 \\ \underline{90} \\ 10 \\ \underline{9} \\ \end{array}$

7. $40X = 0.6$
    $X = 0.6 \div 40 = 0.01\frac{1}{2}$

8. $0.2Y = 0.077$
    $Y = 0.077 \div 0.2 = 0.38\frac{1}{2}$

9. $4.1F = 8.24$
    $F = 8.24 \div 4.1 = 2.00\frac{40}{41}$

10. $11 - 0.99 = 10.01$

11. $5.003 + 21 = 26.003$

12. $100 - 0.01 = 99.99$

13. $\frac{1}{2}(10 \text{ ft})(12 \text{ ft}) = 60 \text{ ft}^2$

14. $\frac{1}{2}(7 \text{ in})(24 \text{ in}) = 84 \text{ in}^2$

15. $\frac{1}{2}(1.3 \text{ m})(4.2 \text{ m}) = 2.73 \text{ m}^2$

16. $\$45.00 \times 1.08 = \$48.60$

17. $\$100.00 \times 3.15 = \$315.00$

18. $3.14(7 \text{ in})^2 = 154 \text{ in}^2$ (rounded)
    $\$10.00 \div 154 \text{ in}^2 = \$0.06 \text{ per in}^2$ (rounded)

19. $3.14(6 \text{ in})^2 \approx 113 \text{ in}^2$
    $\$8.00 \div 113 \text{ in}^2 = \$0.07 \text{ per in}^2$
    14 in is the better buy.

20. $4.25 \text{ cm} \times \frac{10 \text{ mm}}{1 \text{ cm}} = 42.5 \text{ mm}$
    $425 \text{ mm} - 42.5 \text{ mm} = 382.5 \text{ mm}$

Lesson Practice 22A

1. done

2. done

3. $0.09X + 4 = 4.9$
    $0.09X = 4.9 - 4$
    $0.09X = 0.9$
    $X = 0.9 \div 0.09$
    $X = 10$

4. $0.09(10) + 4 = 4.9$
    $0.9 + 4 = 4.9$
    $4.9 = 4.9$

5. $0.6A + 2.8 = 4.66$
    $0.6A = 4.66 - 2.8$
    $0.6A = 1.86$
    $A = 1.86 \div 0.6$
    $A = 3.1$

6. $0.6(3.1) + 2.8 = 4.66$
    $1.86 + 2.8 = 4.66$
    $4.66 = 4.66$

7. $4.6Q + 0.19 = 55.39$
    $4.6Q = 55.39 - 0.19$
    $4.6Q = 55.2$
    $Q = 55.2 \div 4.6$
    $Q = 12$

8. $4.6(12) + 0.19 = 55.39$
   $55.2 + 0.19 = 55.39$
   $55.39 = 55.39$

9. $0.2M + \$10.50 = \$15.50$
   $0.2M = \$15.50 - \$10.50$
   $0.2M = \$5.00$
   $M = \$5.00 \div 0.2$
   $M = \$25.00$

10. $0.10H + 5 \text{ ft} = 13 \text{ ft}$
    $0.10H = 13 \text{ ft} - 5 \text{ ft}$
    $0.10H = 8 \text{ ft}$
    $H = 8 \text{ ft} \div 0.10 = 80 \text{ ft}$

Lesson Practice 22B

1. $1.1W + 2.1 = 5.4$
   $1.1W = 5.4 - 2.1$
   $1.1W = 3.3$
   $W = 3.3 \div 1.1$
   $W = 3$

2. $1.1(3) + 2.1 = 5.4$
   $3.3 + 2.1 = 5.4$
   $5.4 = 5.4$

3. $0.8D + 0.2 = 0.76$
   $0.8D = 0.76 - 0.2$
   $0.8D = 0.56$
   $D = 0.56 \div 0.8$
   $D = 0.7$

4. $0.8(0.7) + 0.2 = 0.76$
   $0.56 + 0.2 = 0.76$
   $0.76 = 0.76$

5. $0.13X + 0.07 = 0.551$
   $0.13X = 0.551 - 0.07$
   $0.13X = 0.481$
   $X = 0.481 \div 0.13$
   $X = 3.7$

6. $0.13(3.7) + 0.07 = 0.551$
   $0.481 + 0.07 = 0.551$
   $0.551 = 0.551$

7. $2.2R + 0.47 = 1.79$
   $2.2R = 1.79 - 0.47$
   $2.2R = 1.32$
   $R = 1.32 \div 2.2$
   $R = 0.6$

8. $2.2(0.6) + 0.47 = 1.79$
   $1.32 + 0.47 = 1.79$
   $1.79 = 1.79$

9. $0.80M + \$14.40 = \$254.40$
   $0.80M = \$254.40 - \$14.40$
   $0.80M = \$240.00$
   $M = \$240.00 \div 0.80$
   $M = \$300.00$

10. $0.10M + \$10.00 = \$15.00$
    $0.10M = \$15.00 - \$10.00$
    $0.10M = \$5.00$
    $M = \$5.00 \div 0.10$
    $M = \$50.00$

Lesson Practice 22C

1. $0.7G + 5.1 = 5.59$
   $0.7G = 5.59 - 5.1$
   $0.7G = 0.49$
   $G = 0.49 \div 0.7$
   $G = 0.7$

2. $0.7(0.7) + 5.1 = 5.59$
   $0.49 + 5.1 = 5.59$
   $5.59 = 5.59$

3. $3.4X + 0.01 = 0.622$
   $3.4X = 0.622 - 0.01$
   $3.4X = 0.612$
   $X = 0.612 \div 3.4$
   $X = 0.18$

4. $3.4(0.18) + 0.01 = 0.622$
   $0.612 + 0.01 = 0.622$
   $0.622 = 0.622$

5. $8.7Y + 1.1 = 5.45$
   $8.7Y = 5.45 - 1.1$
   $8.7Y = 4.35$
   $Y = 4.35 \div 8.7$
   $Y = 0.5$

6. $8.7(0.5) + 1.1 = 5.45$
   $4.35 + 1.1 = 5.45$
   $5.45 = 5.45$

7. $0.01B + 3.3 = 3.341$
   $0.01B = 3.341 - 3.3$
   $0.01B = 0.041$
   $B = 0.041 \div 0.01$
   $B = 4.1$

8. $0.01(4.1)+3.3=3.341$
$0.041+3.3=3.341$
$3.341=3.341$

9. $0.5P+1 \text{ gal}=4 \text{ gal}$
$0.5P=4 \text{ gal}-1 \text{ gal}$
$0.5P=3 \text{ gal}$
$P=3 \text{ gal}\div 0.5$
$P=6 \text{ gal for job}$
$6 \text{ gal}-4 \text{ gal}=2 \text{ gal to buy}$

10. $2.1R+0.6"=32.1"$
$2.1R=32.1"-0.6"$
$2.1R=31.5"$
$R=31.5"\div 2.1$
$R=15"$

Systematic Review 22D

1. $6.8X+0.9=22.66$
$6.8X=22.66-0.9$
$6.8X=21.76$
$X=21.76\div 6.8$
$X=3.2$

2. $6.8(3.2)+0.9=22.66$
$21.76+0.9=22.66$
$22.66=22.66$

3. $5.1Q+4=4.306$
$5.1Q=4.306-4$
$5.1Q=0.306$
$Q=0.306\div 5.1$
$Q=0.06$

4. $5.1(0.06)+4=4.306$
$0.306+4=4.306$
$4.306=4.306$

5. $\begin{array}{r}8.908\approx 8.91\\6\overline{)53.450}\\\underline{48\ 000}\\5\ 450\\\underline{5\ 400}\\50\\\underline{48}\end{array}$

6. $\begin{array}{r}0.271\approx 0.27\\7\overline{)1.900}\\\underline{1\ 400}\\500\\\underline{490}\\10\\\underline{7}\end{array}$

7. $\begin{array}{r}0.204\approx 0.20\\14\overline{)2.860}\\\underline{2\ 800}\\60\\\underline{56}\end{array}$

8. 0.382 hm
9. 0.00017 L
10. 5 dag
11. done
12. $10m\times 10m\times 10m=1{,}000 \text{ m}^3$
13. $1.2 \text{ ft}\times 1.2 \text{ ft}\times 1.2 \text{ ft}=1.728 \text{ ft}^3$
14. $1.7m+0.8m+1.7m+0.8m=5 \text{ m}$
15. $11 \text{ in}+3 \text{ in}+11 \text{ in}+3 \text{ in}=28 \text{ in}$
16. $3.14(6 \text{ ft})=18.84 \text{ ft}$
17. $\$16.64\times 1.15=\$19.136$
($\$19.14$ rounded)
18. $0.5L+12 \text{ ft}^2=132 \text{ ft}^2$
$0.5L=132 \text{ ft}^2-12 \text{ ft}^2$
$0.5L=120 \text{ ft}^2$
$L=120 \text{ ft}^2\div 0.5$
$L=240 \text{ ft}^2$

Systematic Review 22E

1. $0.4T+3.6=12$
$0.4T=12-3.6$
$0.4T=8.4$
$T=8.4\div 0.4$
$T=21$

2. $0.4(21)+3.6=12$
$8.4+3.6=12$
$12=12$

3. $9.9W+0.7=5.749$
$9.9W=5.749-0.7$
$9.9W=5.049$
$W=5.049\div 9.9$
$W=0.51$

4. $9.9(0.51) + 0.7 = 5.749$
   $5.049 + 0.7 = 5.749$
   $5.749 = 5.749$

5. 1.833 or $1.8\overline{3}$
   $$24\overline{)44.000}$$
   $\underline{24\ 000}$
   $20\ 000$
   $\underline{19\ 200}$
   800
   $\underline{720}$
   80
   $\underline{72}$
   8

6. 351.66 or $351.6\overline{6}$
   $$6\overline{)2110.00}$$
   $\underline{1800\ 00}$
   $310\ 00$
   $\underline{300\ 00}$
   $10\ 00$
   $\underline{600}$
   400
   $\underline{360}$
   40
   $\underline{36}$
   4

7. 9.44 or $9.\overline{4}$
   $$9\overline{)85.00}$$
   $\underline{81\ 00}$
   $4\ 00$
   $\underline{3\ 60}$
   40
   $\underline{36}$

8. 0.75 kg
9. 24.5 cm
10. 60 hl
11. 0.03 ft $\times 0.03$ ft $\times 0.03$ ft $= 0.000027$
12. $8m \times 8m \times 8m = 512$ m^3
13. 6.4 in $\times 6.4$ in $\times 6.4$ in $= 262.144$ in^3
14. $1.7m \times 0.8m = 1.36$ m^2
15. 11 in $\times 2$ in $= 22$ in^2
16. $3.14(3\text{ ft})^2 = 28.26$ ft^2
17. $\$48.00 \times 0.55 = \$26.40$
    $\$48.00 - \$26.40 = \$21.60$
    $\$21.60 \times 1.06 = \$22.90$

18. $50,000$ g $= 50$ kg
    50 kg $\times \dfrac{2.2\text{ lb}}{1\text{ kg}} = 110$ lb

19. $\dfrac{5}{9} \times 54$ games
    $54 \div 9 = 6$
    $6 \times 5 = 30$ games won

20. $0.7M + \$5.00 = \$26.49$
    $0.7M = \$26.49 - \$5.00$
    $0.7M = \$21.49$
    $M = \$21.49 \div 0.7 = \$30.70$ needed
    $\$30.70 - \$26.49 = \$4.21$ still to go

Systematic Review 22F

1. $0.08G + 0.59 = 1.39$
   $0.08G = 1.39 - 0.59$
   $0.08G = 0.8$
   $G = 0.8 \div 0.08$
   $G = 10$

2. $0.08(10) + 0.59 = 1.39$
   $0.8 + 0.59 = 1.39$
   $1.39 = 1.39$

3. $0.3X + 2.4 = 2.58$
   $0.3X = 2.58 - 2.4$
   $0.3X = 0.18$
   $X = 0.18 \div 0.3$
   $X = 0.6$

4. $0.3(0.6) + 2.4 = 2.58$
   $0.18 + 2.4 = 2.58$
   $2.58 = 2.58$

5. $0.72\dfrac{6}{7}$
   $$7\overline{)5.10}$$
   $\underline{4\ 90}$
   20
   $\underline{14}$

6. $6.27\dfrac{1}{2}$
   $$2\overline{)12.55}$$
   $\underline{12\ 00}$
   55
   $\underline{40}$
   15
   $\underline{14}$

SYSTEMATIC REVIEW 22F - SYSTEMATIC REVIEW 23D

7. $8 \overline{)12.50} = 1.56\frac{2}{8} = 1.56\frac{1}{4}$
   $\underline{8\ 00}$
   $\ 4\ 50$
   $\ \underline{4\ 00}$
   $\ \ \ \ 50$
   $\ \ \ \ \underline{48}$

8. 6.5 g
9. 0.321 km
10. 62 ml
11. $\frac{5}{2} \text{ft} \times \frac{5}{2} \text{ft} \times \frac{5}{2} \text{ft} = \frac{125}{8} \text{ft}^3 = 15\frac{5}{8} \text{ft}^3$
12. $0.9 \text{m} \times 0.9 \text{m} \times 0.9 \text{m} = 0.729 \text{ m}^3$
13. $4.8 \text{ in} \times 4.8 \text{ in} \times 4.8 \text{ in} = 110.592 \text{ in}^3$
14. $\frac{1}{2}(6 \text{ ft})(16 \text{ ft}) = 48 \text{ ft}^2$
15. $10 \text{ ft} + 10 \text{ ft} + 16 \text{ ft} = 36 \text{ ft}$
16. $\frac{1}{2}(3 \text{ m})(5.1 \text{ m}) = 7.65 \text{ m}^2$
17. $4.5 \text{m} + 5.1 \text{m} + 3.3 \text{m} = 12.9 \text{ m}$
18. $3 \text{ ft} \times 3 \text{ ft} = 9 \text{ ft}^2$
19. $12 \text{ ft} \times 15 \text{ ft} = 180 \text{ ft}^2$
    $180 \text{ ft}^2 \div \frac{9 \text{ ft}^2}{1 \text{ yd}^2} = 20 \text{ yd}^2$
20. $20 \text{ yd}^2 \times (\$9.99/\text{yd}^2) = \$199.80$

Lesson Practice 23A

1. done
2. $6 \div 7 = 0.85\frac{5}{7} = 85\frac{5}{7}\%$
3. $3 \div 10 = 0.30 = 30\%$
4. $2 \div 5 = 0.40 = 40\%$
5. $7 \div 8 = 0.87\frac{1}{2} = 87\frac{1}{2}\%$
6. $10 \div 11 = 0.90\frac{10}{11} = 90\frac{10}{11}\%$
7. $\frac{13}{15} = 13 \div 15 = 0.86\frac{2}{3} = 86\frac{2}{3}\%$
8. $\frac{1}{6} = 1 \div 6 = 0.16\frac{2}{3} = 16\frac{2}{3}\%$
9. $\frac{16}{50} = 16 \div 50 = 0.32 = 32\%$

10. $50 - 16 = 34$ not wearing hats
    $\frac{34}{50} = 34 \div 50 = 0.68 = 68\%$
    (or $100\% - 32\% = 68\%$)

Lesson Practice 23B

1. $7 \div 10 = 0.70 = 70\%$
2. $1 \div 2 = 0.50 = 50\%$
3. $4 \div 9 = 0.44\frac{4}{9} = 44\frac{4}{9}\%$
4. $3 \div 11 = 0.27\frac{3}{11} = 27\frac{3}{11}\%$
5. $3 \div 5 = 0.60 = 60\%$
6. $1 \div 12 = 0.08\frac{1}{3} = 8\frac{1}{3}\%$
7. $3 \div 4 = 0.75 = 75\%$
8. $5 \div 6 = 0.83\frac{1}{3} = 83\frac{1}{3}\%$
9. $25 - 3 = 22$ correct
   $22 \div 25 = 0.88 = 88\%$
10. $35 \div 100 = 0.35 = 35\%$

Lesson Practice 23C

1. $1 \div 7 = 0.14\frac{2}{7} = 14\frac{2}{7}\%$
2. $2 \div 9 = 0.22\frac{2}{9} = 22\frac{2}{9}\%$
3. $4 \div 5 = 0.80 = 80\%$
4. $5 \div 12 = 0.41\frac{2}{3} = 41\frac{2}{3}\%$
5. $2 \div 3 = 0.66\frac{2}{3} = 66\frac{2}{3}\%$
6. $3 \div 8 = 0.37\frac{1}{2} = 37\frac{1}{2}\%$
7. $87 \div 100 = 0.87 = 87\%$
8. $5 \div 8 = 0.62\frac{1}{2} = 62\frac{1}{2}\%$
9. $1 \div 3 = 0.33\frac{1}{3} = 33\frac{1}{3}\%$
10. $1 \div 2 = 0.50 = 50\%$

Systematic Review 23D

1. $1 \div 4 = 0.25 = 25\%$
2. $7 \div 9 = 0.77\frac{7}{9} = 77\frac{7}{9}\%$

3. $5 \div 16 = 0.31\frac{1}{4} = 31\frac{1}{4}\%$

4. $1 \div 8 = 0.12\frac{1}{2} = 12\frac{1}{2}\%$

5. $0.17X + 1.2 = 1.234$
$0.17X = 1.234 - 1.2$
$0.17X = 0.034$
$X = 0.034 \div 0.17$
$X = 0.2$

6. $0.17(0.2) + 1.2 = 1.234$
$0.034 + 1.2 = 1.234$
$1.234 = 1.234$

7. $2.046 \approx 2.05$
```
13 ) 26.600
     26 000
        600
        520
         80
         78
```

8. $0.822 \approx 0.82$
```
 9 ) 7.400
     7 200
       200
       180
        20
        18
```

9. $0.198 \approx 0.20$
```
23 ) 4.570
     2 300
     2 270
     2 070
       200
       184
```

10. done

11. $12m \times 4m \times 4.1m = 196.8 \ m^3$

12. $2.3 \ ft \times 1.8 \ ft \times 3.2 \ ft = 13.248 \ ft^3$

13. $38 \div 45 = 0.84\frac{4}{9} = 84\frac{4}{9}\%$

14. $6 \div 8 = 0.75 = 75\%$

15. $4mi \times 2mi \times 2mi = 16 \ mi^3$

16. $12in \times 12in = 144 \ in^2$

17. $10 \ ft \times 10 \ ft = 100 \ ft^2$
$100 \ ft^2 \times 144 \ in^2/ft^2 = 14,400 \ in^2$

18. $\$15.99 \times 0.40 = \$6.396$ or
$\$6.40$ off the regular price

Systematic Review 23E

1. $1 \div 9 = 0.11\frac{1}{9} = 11\frac{1}{9}\%$

2. $2 \div 3 = 0.66\frac{2}{3} = 66\frac{2}{3}\%$

3. $11 \div 14 = 0.78\frac{4}{7} = 78\frac{4}{7}\%$

4. $4 \div 7 = 0.57\frac{1}{7} = 57\frac{1}{7}\%$

5. $0.3Y + 0.09 = 1.44$
$0.3Y = 1.44 - 0.09$
$0.3Y = 1.35$
$Y = 1.35 \div 0.3$
$Y = 4.5$

6. $0.3(4.5) + 0.09 = 1.44$
$1.35 + 0.09 = 1.44$
$1.44 = 1.44$

7. 8.909 or $8.\overline{90}$
```
11 ) 98.000
     88 000
     10 000
      9 900
        100
         99
```

8. 12.9833 or $12.98\overline{3}$
```
 6 ) 77.9000
     60 0000
     17 0000
     12 0000
      5 0000
        5000
        4800
         200
         180
```

SYSTEMATIC REVIEW 23E - SYSTEMATIC REVIEW 23F

9. $\phantom{48\overline{)}}104.166$ or $104.1\overline{6}$
    $48\overline{)5000.000}$
    $\phantom{48\overline{)}}\underline{4800\ 000}$
    $\phantom{48\overline{)00}}200\ 000$
    $\phantom{48\overline{)00}}\underline{192\ 000}$
    $\phantom{48\overline{)0000}}8\ 000$
    $\phantom{48\overline{)0000}}\underline{4\ 800}$
    $\phantom{48\overline{)0000}}3\ 200$
    $\phantom{48\overline{)0000}}\underline{2\ 880}$
    $\phantom{48\overline{)000000}}320$
    $\phantom{48\overline{)000000}}\underline{288}$

10. $7in \times 6in \times 5in = 210\ in^3$
11. $7.5m \times 2.4m \times 2m = 36\ m^3$
12. $\frac{7}{2}ft \times \frac{9}{4}ft \times \frac{29}{8}ft = \frac{1827}{64} = 28\frac{35}{64}\ ft^3$
13. $102 + 60 = 162$ total games
    $102 \div 162 = 0.62\frac{26}{27} = 62\frac{26}{27}\%$
14. $60 \div 162 = 0.37\frac{1}{27} = 37\frac{1}{27}\%$
    $0.37\frac{1}{27} + 62\frac{26}{27} = 0.99\frac{27}{27} = 100\%$
15. $\$47.50 \times 1.08 = \$51.30$
16. round pizza:
    $3.14(6\ in)^2 = 113\ in^2$
    $\$5.65 \div 113\ in^2 = \$0.05$ per in^2
    rectangular pizza:
    $12\ in \times 24\ in = 288\ in^2$
    $\$11.52 \div 288\ in^2 = \$0.04$ per in^2
    Rectangular pizza has better price.
17. $288\ in^2 \times \frac{1\ ft^2}{144\ in^2} = 2\ ft^2$
    or $12\ in \times 24\ in = 1\ ft \times 2\ ft = 2\ ft^2$
18. $\$4.20 \times 0.75 = \$3.15$
19. $36in + 42in + 36in + 42in = 156\ in$
20. $156\ ft \div 1\ ft/12\ in = 13\ ft$

Systematic Review 23F

1. $7 \div 15 = 0.46\frac{2}{3} = 46\frac{2}{3}\%$
2. $3 \div 7 = 0.42\frac{6}{7} = 42\frac{6}{7}\%$
3. $5 \div 9 = 0.55\frac{5}{9} = 55\frac{5}{9}\%$
4. $1 \div 10 = 0.10 = 10\%$
5. $6.7R + 0.4 = 6.43$
   $6.7R = 6.43 - 0.4$
   $6.7R = 6.03$
   $R = 6.03 \div 6.7$
   $R = 0.9$
6. $6.7(0.9) + 0.4 = 6.43$
   $6.03 + 0.4 = 6.43$
   $6.43 = 6.43$
7. $\phantom{33\overline{)}}1.56\frac{2}{33}$
   $33\overline{)51.50}$
   $\phantom{33\overline{)}}\underline{3300}$
   $\phantom{33\overline{)}}1850$
   $\phantom{33\overline{)}}\underline{1650}$
   $\phantom{33\overline{)00}}200$
   $\phantom{33\overline{)00}}\underline{198}$
   $\phantom{33\overline{)000}}2$
8. $\phantom{8\overline{)}}8.23\frac{6}{8} = 8.23\frac{3}{4}$
   $8\overline{)65.90}$
   $\phantom{8\overline{)}}\underline{6400}$
   $\phantom{8\overline{)0}}190$
   $\phantom{8\overline{)0}}\underline{160}$
   $\phantom{8\overline{)00}}30$
   $\phantom{8\overline{)00}}\underline{24}$
   $\phantom{8\overline{)000}}6$
9. $\phantom{06\overline{)}}11.96\frac{4}{6} = 11.96\frac{2}{3}$
   $06\overline{)71.80}$
   $\phantom{06\overline{)}}\underline{6000}$
   $\phantom{06\overline{)}}1180$
   $\phantom{06\overline{)0}}\underline{600}$
   $\phantom{06\overline{)0}}580$
   $\phantom{06\overline{)0}}\underline{540}$
   $\phantom{06\overline{)00}}40$
   $\phantom{06\overline{)00}}\underline{36}$
   $\phantom{06\overline{)000}}4$
10. $80in \times 90in \times 60in = 432{,}000\ in^3$
11. $13.1m \times 4.6m \times 2.9m = 174.754\ m^3$
12. $0.09ft \times 0.05ft \times 1ft = 0.0045\ ft^3$
13. $3ft \times 3ft \times 3ft = 27\ ft^3$
14. $27 \times 2\frac{1}{3} = \frac{\overset{9}{\cancel{27}}}{1} \times \frac{7}{\cancel{3}} = \frac{63}{1} = 63\ ft^3$

15. 6 yd³ × 27 ft³/yd³ = 162 ft³
16. 6 × $55.50 = $333
17. 6yd³ ÷ 8yd³ = 0.75 = 75%
18. 34 ÷ 8 = 4.25, so 5 trips
19. $32 × 0.30 = $9.60
    $32 − $9.60 = $22.40 before tax
    $22.40 × 1.04 = $23.296 or $23.30
    $25 − $23.30 = $1.70
20. $7.8H = 27.3$
    $\frac{7.8}{7.8}H = \frac{27.3}{7.8}$
    27.3 ft² ÷ 7.8 ft = 3.5 ft

Lesson Practice 24A

1. done
2. $\frac{96}{100} = \frac{24}{25}$
3. $\frac{982}{1000} = \frac{491}{500}$
4. $\frac{6}{1000} = \frac{3}{500}$
5. $\frac{18}{1000} = \frac{9}{500}$
6. $\frac{9}{10}$
7. $\frac{885}{1000} = \frac{177}{200}$
8. $\frac{84}{100} = \frac{21}{25}$
9. $\frac{32}{100} = \frac{8}{25}$
10. $\frac{15}{1000} = \frac{3}{200}$
11. $\frac{20}{100} = \frac{1}{5}$; 5 pieces
12. $\frac{45}{100} = \frac{9}{20}$; 9 people

Lesson Practice 24B

1. $\frac{915}{1000} = \frac{183}{200}$
2. $\frac{53}{1000}$
3. $\frac{7}{1000}$

4. $\frac{62}{100} = \frac{31}{50}$
5. $\frac{75}{100} = \frac{3}{4}$
6. $\frac{52}{100} = \frac{13}{25}$
7. $\frac{28}{1000} = \frac{7}{250}$
8. $\frac{725}{1000} = \frac{29}{40}$
9. $\frac{6}{10} = \frac{3}{5}$
10. $\frac{95}{100} = \frac{19}{20}$
11. $\frac{4}{10} = \frac{2}{5}$; 2 juncos
12. $\frac{125}{1000} = \frac{1}{8}$; 8 pieces

Lesson Practice 24C

1. $\frac{68}{100} = \frac{17}{25}$
2. $\frac{25}{1000} = \frac{1}{40}$
3. $\frac{84}{100} = \frac{21}{25}$
4. $\frac{16}{1000} = \frac{2}{125}$
5. $\frac{8}{1000} = \frac{1}{125}$
6. $\frac{7}{10}$
7. $\frac{325}{1000} = \frac{13}{40}$
8. $\frac{743}{1000}$
9. $\frac{56}{100} = \frac{14}{25}$
10. $\frac{11}{100}$
11. $\frac{4}{100} = \frac{1}{25}$; 25 pieces
12. 0.75 × 4 = 3 houses

Systematic Review 24D

1. $\dfrac{25}{1000} = \dfrac{1}{40}$
2. $\dfrac{18}{100} = \dfrac{9}{50}$
3. $1 \div 5 = 0.20 = 20\%$
4. $5 \div 8 = 0.62\dfrac{1}{2} = 62\dfrac{1}{2}\%$
5. $0.19X + 0.73 = 0.825$
   $0.19X = 0.825 - 0.73$
   $0.19X = 0.095$
   $X = 0.095 \div 0.19$
   $X = 0.5$
6. $0.19(0.5) + 0.73 = 0.825$
   $0.095 + 0.73 = 0.825$
   $0.825 = 0.825$
7. $\phantom{15\,\overline{\smash{)}}}\ 2.306 = 2.31$
   $15\,\overline{)34.600}$
   $\phantom{15\,\overline{)}}\underline{30000}$
   $\phantom{15\,\overline{)0}}4600$
   $\phantom{15\,\overline{)0}}\underline{4500}$
   $\phantom{15\,\overline{)000}}100$
   $\phantom{15\,\overline{)0000}}\underline{90}$
   $\phantom{15\,\overline{)0000}}10$
8. $\phantom{8\,\overline{)}}0.837 = 0.84$
   $8\,\overline{)6.700}$
   $\phantom{8\,\overline{)}}\underline{6400}$
   $\phantom{8\,\overline{)0}}300$
   $\phantom{8\,\overline{)0}}\underline{240}$
   $\phantom{8\,\overline{)00}}60$
   $\phantom{8\,\overline{)00}}\underline{56}$
   $\phantom{8\,\overline{)000}}4$
9. $\phantom{34\,\overline{)}}0.036 = 0.04$
   $34\,\overline{)1.230}$
   $\phantom{34\,\overline{)}}\underline{1020}$
   $\phantom{34\,\overline{)0}}210$
   $\phantom{34\,\overline{)0}}\underline{204}$
   $\phantom{34\,\overline{)00}}6$
10. $2 + 7 + 9 = 18$
    $18 \div 3 = 6$
11. $5 + 5 + 9 + 13 = 32$
    $32 \div 4 = 8$
12. $2 + 5 + 8 + 9 = 24$
    $24 \div 4 = 6$
13. $77 + 80 + 95 + 100 = 352$
    $352 \div 4 = 88$
14. $19.06 \text{ ft/sec} \times 60 \text{ sec} = 1{,}143.6 \text{ ft}$
15. $1.15 \times 5 \text{ hr} = 5.75 \text{ hr}$
16. $\dfrac{1}{2}(0.24\text{m} \times 0.5\text{m}) = 0.06 \text{ m}^2$
17. $2{,}500 \text{ ml} \times \dfrac{1 \text{ L}}{1{,}000 \text{ ml}} = 2.5 \text{ liters}$
18. $4 \text{ km} \times 10 \text{ hm/km} = 40 \text{ hm}$

Systematic Review 24E

1. $\dfrac{625}{1000} = \dfrac{5}{8}$
2. $\dfrac{21}{100}$
3. $2 \div 7 = 0.28\dfrac{4}{7} = 28\dfrac{4}{7}\%$
4. $1 \div 4 - 0.25 = 25\%$
5. $0.3X + 9.1 = 9.244$
   $0.3X = 9.244 - 9.1$
   $0.3X = 0.144$
   $X = 0.144 \div 0.3$
   $X = 0.48$
6. $0.3(0.48) + 9.1 = 9.244$
   $0.144 + 9.1 = 9.244$
   $9.244 = 9.244$
7. $\phantom{9\,\overline{)}}9.044 = 9.0\overline{4}$
   $9\,\overline{)81.400}$
   $\phantom{9\,\overline{)}}\underline{81000}$
   $\phantom{9\,\overline{)00}}400$
   $\phantom{9\,\overline{)00}}\underline{360}$
   $\phantom{9\,\overline{)000}}40$
   $\phantom{9\,\overline{)000}}\underline{36}$
   $\phantom{9\,\overline{)0000}}4$
8. $\phantom{3\,\overline{)}}48.66 = 48.\overline{6}$
   $3\,\overline{)146.00}$
   $\phantom{3\,\overline{)}}\underline{12000}$
   $\phantom{3\,\overline{)0}}2600$
   $\phantom{3\,\overline{)0}}\underline{2400}$
   $\phantom{3\,\overline{)00}}200$
   $\phantom{3\,\overline{)00}}\underline{180}$
   $\phantom{3\,\overline{)000}}20$
   $\phantom{3\,\overline{)000}}\underline{18}$
   $\phantom{3\,\overline{)0000}}2$

SYSTEMATIC REVIEW 24E - SYSTEMATIC REVIEW 24F

9. $4.8181 = 4.\overline{81}$
    $11\overline{)53.0000}$
    $\underline{44 0000}$
    $9 0000$
    $\underline{8 8000}$
    2000
    $\underline{1100}$
    900
    $\underline{880}$
    20

10. $4+5+6=15$
    $15 \div 3 = 5$

11. $5+8+8=21$
    $21 \div 3 = 7$

12. $6+7+9+10=32$
    $32 \div 4 = 8$

13. $76+81+89+92=338$
    $338 \div 4 = 84.5$

14. $10m \times 10m \times 10m = 1{,}000 \text{ m}^3$

15. $6 \text{ in} + 9 \text{ in} + 11 \text{ in} = 26 \text{ in}$

16. $10{,}000{,}000 \text{ m} \times \dfrac{1 \text{ km}}{1{,}000 \text{ m}} = 10{,}000 \text{ km}$

17. $384 \text{ oz} \times 0.75 = 288 \text{ oz}$

18. $288 \text{ oz} \times \dfrac{1 \text{ lb}}{16 \text{ oz}} = 18 \text{ lb}$

19. $\$9.00 \times 1.10 = \$9.90$

20. $3.14(4 \text{ in})^2 = 50.24 \text{ in}^2$

Systematic Review 24F

1. $\dfrac{92}{100} = \dfrac{23}{25}$

2. $\dfrac{45}{1000} = \dfrac{9}{200}$

3. $1 \div 9 = 0.11\dfrac{1}{9} = 11\dfrac{1}{9}\%$

4. $3 \div 5 = 0.60 = 60\%$

5. $0.2X + 75 = 75.18$
   $0.2X = 75.18 - 75$
   $0.2X = 0.18$
   $X = 0.18 \div 0.2$
   $X = 0.9$

6. $0.2(0.9) + 75 = 75.18$
   $0.18 + 75 = 75.18$
   $75.18 = 75.18$

7. $166.66\dfrac{4}{6} = 166.66\dfrac{2}{3}$
    $6\overline{)1000.00}$
    $\underline{600 00}$
    $400 00$
    $\underline{360 00}$
    4000
    $\underline{3600}$
    400
    $\underline{360}$
    40
    $\underline{36}$
    4

8. $0.32\dfrac{1}{2}$
    $2\overline{)0.65}$
    $\underline{6 0}$
    5
    $\underline{4}$
    1

9. $2.11\dfrac{3}{17}$
    $17\overline{)35.90}$
    $\underline{34 00}$
    190
    $\underline{170}$
    20
    $\underline{17}$
    3

10. $3+6+9=18$
    $18 \div 3 = 6$

11. $4+8+10+14=36$
    $36 \div 4 = 9$

12. $1+3+6+10=20$
    $20 \div 4 = 5$

13. $21 \text{ in} + 34 \text{ in} + 21 \text{ in} + 34 \text{ in} = 110 \text{ in}$

14. $(2)(3.14)(4 \text{ ft}) = 25.12 \text{ ft}$

15. $\$2{,}500 \times 1.17 = \$2{,}925$

16. $250 \div 100 = 2.5$; 3 packs

17. $\$1.29 \times 3 = \$3.87$
    $\$3.87 \times 1.05 \approx \$4.06$

18. $3.25 \times 40 \text{ lb} = 130 \text{ lb}$

19. $130 \text{ lb} \times \dfrac{1 \text{ bag}}{10 \text{ lb}} = 13 \text{ bags}$

20. $13 \times \$1.45 = \$18.85$

Lesson Practice 25A

1. done
2. $\dfrac{3+8+8+8+9+9+11}{7} = 8$
   mean = 8
   median = 8
   mode = 8
3. $\dfrac{3+8+9+9+11}{5} = 8$
   mean = 8
   median = 9
   mode = 9
4. $\dfrac{3+4+6+6+6+8+9}{7} = 6$
   mean = 6
   median = 6
   mode = 6
5. $\dfrac{3+8+8+10+12+19}{6} = 10$
   mean = 10
   median = (8+10)/2 = 9
   mode = 8
6. $\dfrac{11+12+15+15+33+34}{6} = 20$
   mean = 20
   median = 15
   mode = 15
7. 6
8. 29° average temperature
9. 3

Lesson Practice 25B

1. $\dfrac{10+11+15+15+15+16+23}{7} = 15$
   mean = 15
   median = 15
   mode = 15
2. $\dfrac{1+1+1+1+2+2+2+3+3+4}{10} = 2$
   mean = 2
   median = 2
   mode = 1
3. $\dfrac{3+4+7+7+7+9+19}{7} = 8$
   mean = 8
   median = 7
   mode = 7
4. $\dfrac{13+14+24+24+25}{5} = 20$
   mean = 20
   median = 24
   mode = 24
5. $\dfrac{15+17+17+17+19+23}{6} = 18$
   mean = 18
   median = 17
   mode = 17
6. $\dfrac{9+10+10+10+10+11+13+13+13}{9} = 11$
   mean = 11
   median = 10
   mode = 10
7. 17° + 30° + 32° + 35° + 40° = 154°
   154° ÷ 5 = 30.8°
8. 98
9. median

Lesson Practice 25C

1. $\dfrac{5+8+8+9+15}{5} = 9$
   mean = 9
   median = 8
   mode = 8
2. $\dfrac{9+11+11+14+20+23+24}{7} = 16$
   mean = 16
   median = 14
   mode = 11
3. $\dfrac{7+7+7+7+14+18}{6} = 10$
   mean = 10
   median = 7
   mode = 7
4. $\dfrac{1+6+6+19}{4} = 8$
   mean = 8
   median = 6
   mode = 6

LESSON PRACTICE 25C - SYSTEMATIC REVIEW 25E

5. $\dfrac{11+13+13+13+19+21}{6} = 15$

    mean = 15
    median = 13
    mode = 13

6. $\dfrac{8+9+11+14+15+15+19}{7} = 13$

    mean = 13
    median = 14
    mode = 15

7. $5+6+7+8+9 = 35$
    $35 \div 5 = 7$

8. 5'10"; median

9. 8; mode

Systematic Review 25D

1. $\dfrac{5+6+6+9+10+10+10}{7} = 8$

    mean = 8
    median = 9
    mode = 10

2. $\dfrac{7+8+11+11+13}{5} = 10$

    mean = 10
    median = 11
    mode = 11

3. $\dfrac{13}{100}$

4. $\dfrac{350}{1000} = \dfrac{35}{100} = \dfrac{7}{20}$

5. $2 \div 9 = 0.22\dfrac{2}{9} = 22\dfrac{2}{9}\%$

6. $1 \div 10 = 0.10 = 10\%$

7. $0.7X + 1.5 = 3.88$
    $ 0.7X = 3.88 - 1.5$
    $ 0.7X = 2.38$
    $ X = 2.38 \div 0.7$
    $ X = 3.4$

8. $0.7(3.4) + 1.5 = 3.88$
    $2.38 + 1.5 = 3.88$
    $3.88 = 3.88$

9. $\pi(4.5 \text{ in})^2 \approx 63.59 \text{ in}^2$

10. $\pi(9 \text{ in}) \approx 28.26 \text{ in}$

11. $8 \times 8 = 64$

12. $3 \times 3 = 9$

13. $10 \times 10 = 100$

14. $15 \text{ ft} + 15 \text{ ft} + 24 \text{ ft} = 54 \text{ ft}$

15. $12 \text{ m} + 16 \text{ m} + 20 \text{ m} = 48 \text{ m}$

16. $1.5 \text{ ft} + 2.5 \text{ ft} + 2 \text{ ft} = 6 \text{ ft}$

17. $\dfrac{1}{2}(24 \text{ ft} \times 9 \text{ ft}) = 108 \text{ ft}^2$

18. $\dfrac{1}{2}(12 \text{ m} \times 16 \text{ m}) = 96 \text{ m}^2$

19. $\dfrac{1}{2}(1.2 \text{ ft} \times 2.5 \text{ ft}) = 1.5 \text{ ft}^2$

20. $50 \text{ g} \times 1{,}000 \text{ mg/g} = 50{,}000 \text{ mg total}$
    $50{,}000 \text{ mg} \div 100 \text{ tablets} = 500 \text{ mg/tablet}$

Systematic Review 25E

1. $\dfrac{2+4+4+4+7+9}{6} = 5$

    mean = 5
    median = 4
    mode = 4

2. $\dfrac{3+8+8+15+16+29+33}{7} = 16$

    mean = 16
    median = 15
    mode = 8

3. $\dfrac{8}{10} = \dfrac{4}{5}$

4. $\dfrac{48}{100} = \dfrac{12}{25}$

5. $5 \div 6 = 0.83\dfrac{1}{3} = 83\dfrac{1}{3}\%$

6. $3 \div 15 = 0.2 = 20\%$

7. $1.1X + 2.2 = 9.9$
    $1.1X = 9.9 - 2.2$
    $1.1X = 7.7$
    $X = 7.7 \div 1.1$
    $X = 7$

8. $1.1(7) + 2.2 = 9.9$
    $7.7 + 2.2 = 9.9$
    $9.9 = 9.9$

9. $3.14(5 \text{ in})^2 = 78.5 \text{ in}^2$

10. $3.14(10 \text{ in}) = 31.4 \text{ in}$

11. $2 \times 2 \times 2 = 8$

12. $1 \times 1 \times 1 \times 1 \times 1 = 1$

13. $7 \times 7 = 49$

SYSTEMATIC REVIEW 25E - LESSON PRACTICE 26B

14. $2.6 \text{ ft} + 2.6 \text{ ft} + 2.6 \text{ ft} + 2.6 \text{ ft} = 10.4 \text{ ft}$
15. $0.1 \text{ m} + 0.51 \text{ m} + 0.1 \text{ m} + 0.51 \text{ m} = 1.22 \text{ m}$
16. $3.1 \text{ ft} + 5.4 \text{ ft} + 3.1 \text{ ft} + 5.4 \text{ ft} = 17 \text{ ft}$
17. $2.6 \text{ ft} \times 2.6 \text{ ft} = 6.76 \text{ ft}^2$
18. $0.1 \text{ m} \times 0.51 \text{ m} = 0.051 \text{ m}^2$
19. $3.1 \text{ ft} \times 5.4 \text{ ft} = 16.74 \text{ ft}^2$
20. $30,300 \text{ L} \div 1,000 \text{ L/kl} = 30.3 \text{ kl}$

Systematic Review 25F

1. $\frac{1+1+1+1+1+5+6+8}{8} = 3$
   mean = 3
   median = 1
   mode = 1
2. $\frac{5+9+9+9+13+15}{6} = 10$
   mean = 10
   median = 9
   mode = 9
3. $\frac{8}{100} = \frac{2}{25}$
4. $\frac{175}{1000} = \frac{7}{40}$
5. $9 \div 10 = 0.9 = 90\%$
6. $7 \div 8 = 0.87\frac{1}{2} = 87\frac{1}{2}\%$
7. $8X + 0.09 = 1.69$
   $8X = 1.69 - 0.09$
   $8X = 1.6$
   $X = 1.6 \div 8$
   $X = 0.2$
8. $8(0.2) + 0.09 = 1.69$
   $1.6 + 0.09 = 1.69$
   $1.69 = 1.69$
9. $3.14(3.1 \text{ ft})^2 = 30.1754 \text{ ft}^2$
10. $3.14(6.2 \text{ ft}) = 19.468 \text{ ft}$
11. $4 \times 4 = 16$
12. $6 \times 6 = 36$
13. $5 \times 5 \times 5 = 125$
14. $4.2 \text{ m} + 5 \text{ m} + 4.2 \text{ m} + 5 \text{ m} = 18.4 \text{ m}$
15. $18 \text{ mm} + 9 \text{ mm} + 18 \text{ mm} + 9 \text{ mm} = 54 \text{ mm}$
16. $2.4 \text{ in} + 1.8 \text{ in} + 2.4 \text{ in} + 1.8 \text{ in} = 8.4 \text{ in}$
17. $4.2 \text{ m} \times 4 \text{ m} = 16.8 \text{ m}^2$
18. $18 \text{ mm} \times 6 \text{ mm} = 108 \text{ mm}^2$
19. $2.4 \text{ in} \times 1.2 \text{ in} = 2.88 \text{ in}^2$
20. $3 \text{ dm} \times \frac{10 \text{ cm}}{1 \text{ dm}} = 30$ one-cm pieces

Lesson Practice 26A

1. done
2. done
3. done
4. $\frac{7}{50}$
5. $\frac{20}{300} = \frac{1}{15}$
6. $\frac{240}{300} = \frac{4}{5}$
7. $\frac{40}{300} = \frac{2}{15}$
8. $\frac{300-20}{300} = \frac{280}{300} = \frac{14}{15}$
9. $\frac{2}{1000} = \frac{1}{500}$
10. $\frac{10}{1000} = \frac{1}{100}$
11. $\frac{5}{500} = \frac{1}{100}$
12. $\frac{32}{192+32} = \frac{32}{224} = \frac{1}{7}$

Lesson Practice 26B

1. $\frac{25}{80} = \frac{5}{16}$
2. $\frac{20+15}{80} = \frac{35}{80} = \frac{7}{16}$
3. $\frac{20+25+15}{80} = \frac{60}{80} = \frac{3}{4}$
4. $\frac{20+25+13}{80} = \frac{58}{80} = \frac{29}{40}$
5. $\frac{10}{120} = \frac{1}{12}$
6. $\frac{100}{120} = \frac{5}{6}$
7. $\frac{10}{120} = \frac{1}{12}$
8. $\frac{100+10}{120} = \frac{110}{120} = \frac{11}{12}$
9. $\frac{4}{10} = \frac{2}{5}$

10. $\frac{7}{10}$

11. $\frac{4}{10} = \frac{2}{5}$

12. $\frac{3}{10}$

Lesson Practice 26C

1. $\frac{30}{250} = \frac{3}{25}$

2. $\frac{50+40+30+30}{250} = \frac{150}{250} = \frac{3}{5}$

3. $\frac{40+30}{250} = \frac{70}{250} = \frac{7}{25}$

4. $\frac{100+50+30}{250} = \frac{180}{250} = \frac{18}{25}$

5. $\frac{10}{3500} = \frac{1}{350}$

6. $\frac{500}{3500} = \frac{1}{7}$

7. $\frac{3500-500}{3500} = \frac{3000}{3500} = \frac{6}{7}$

8. $\frac{4}{10} = \frac{2}{5}$

9. $\frac{7}{10}$

10. $\frac{1}{10}$

11. $\frac{2}{10} = \frac{1}{5}$

12. $\frac{568 \div 2}{568} = \frac{284}{568} = \frac{1}{2}$

Systematic Review 26D

1. $\frac{60}{540} = \frac{1}{9}$

2. $\frac{450}{540} = \frac{5}{6}$

3. $\frac{450+60}{540} = \frac{510}{540} = \frac{17}{18}$

4. $\frac{5+9+9+9+13+15}{6} = 10$

   mean = 10
   median = 9
   mode = 9

5. $\frac{1+1+3+3+3+3+5+6+6+9}{10} = 4$

   mean = 4
   median = 3
   mode = 3

6. $\frac{34}{100} = \frac{17}{50}$

7. $\frac{178}{1000} = \frac{89}{500}$

8. $0.75 = 75\%$

9. $0.55\frac{5}{9} = 55\frac{5}{9}\%$

10. $5+4=9$; $9 \times 8 = 72$; $72-2=70$;
    $70 \div 10 = 7$; $7+3 = 10$

11. $7-3=4$; $4 \times 6 = 24$; $24 \div 3 = 8$;
    $8 \times 9 = 72$

12. $\frac{2}{5000} = \frac{1}{2500}$

13. 45 lb ÷ 0.9 lb / customer = 50 customers

14. $\$48.00 \times 0.20 = \$9.60$
    $\$48.00 - \$9.60 = \$38.40$

Systematic Review 26E

1. $\frac{100}{500} = \frac{1}{5}$

2. $\frac{250+100}{500} = \frac{350}{500} = \frac{7}{10}$

3. $\frac{30+40+100}{500} = \frac{170}{500} = \frac{17}{50}$

4. $\frac{7+8+10+10+14+17}{6} = 11$

   mean = 11
   median = 10
   mode = 10

5. $\frac{15+21+22+27+27}{5} = 22.4$

   mean = 22.4
   median = 22
   mode = 27

6. $\frac{48}{100} = \frac{12}{25}$

7. $\frac{525}{1000} = \frac{21}{40}$

8. $0.83\frac{1}{3} = 83\frac{1}{3}\%$

9. $0.91\frac{2}{3} = 91\frac{2}{3}\%$

10. $36,320 \text{ g} \times \dfrac{1 \text{ kg}}{1,000 \text{ g}} = 36.32 \text{ kg}$

11. $\pi(10 \text{ ft})^2 \approx 314 \text{ ft}^2$

12. $25.02 \text{ gal} + 20.6 \text{ gal} + 0.25 \text{ gal} = 45.87 \text{ gal}$

13. $125.5 \text{ ft} \times 200 \text{ ft} = 25,100 \text{ ft}^2$

14. $9 + 4 = 13$; $13 + 3 = 16$; $16 + 4 = 20$;
    $20 \div 5 = 4$; $4 - 0 = 4$

15. $10 \div 2 = 5$; $5 + 2 = 7$; $7 \times 7 = 49$;
    $49 + 1 = 50$; $50 - 20 = 30$

Systematic Review 26F

1. $\dfrac{2}{10} = \dfrac{1}{5}$

2. $\dfrac{9}{10}$

3. $\dfrac{4}{10} = \dfrac{2}{5}$

4. $\dfrac{17 + 17 + 19 + 22 + 25}{5} = 20$

    mean = 20
    median = 19
    mode = 17

5. $\dfrac{1 + 1 + 1 + 7 + 7 + 8 + 9 + 14}{8} = 6$

    mean = 6
    median = 7
    mode = 1

6. $\dfrac{65}{100} = \dfrac{13}{20}$

7. $\dfrac{125}{1000} = \dfrac{1}{8}$

8. $0.42\dfrac{6}{7} = 42\dfrac{6}{7}\%$

9. $0.90 = 90\%$

10. $200 \times 0.87 = 174$

11. $1 \text{ km} \times \dfrac{1,000 \text{ m}}{1 \text{ km}} \times \dfrac{1 \text{ hm}}{100 \text{ m}} = 10 \text{ hm}$;
    10 runners

12. $3.14(12 \text{ ft}) = 37.68 \text{ ft}$; 38 lengths

13. meter

14. $\dfrac{1}{2} \times 8 = 4$; $4 + 3 = 7$;
    $7 \times 7 = 49$; $49 + 1 = 50$

15. $8 + 9 = 17$; $17 - 7 = 10$;
    $10 \div 5 = 2$; $2 \times 4 = 8$

Lesson Practice 27A

1. ray
2. line segment
3. geometry
4. point
5. line
6. d: line
7. c: ray
8. b: line segment
9. e: point
10. a: infinity
11. A
12. $\overleftrightarrow{ST}$ or $\overleftrightarrow{TS}$

Lesson Practice 27B

1. ray
2. line segment
3. endpoint
4. points
5. point
6. d
7. a
8. c
9. b
10. $\overline{LM}$ or $\overline{ML}$
11. $\overrightarrow{QR}$
12. line

Lesson Practice 27C

1. point
2. line
3. ray
4. line segment
5. ray
6. e
7. d
8. a
9. c
10. b
11. $\overleftrightarrow{DE}$ or $\overleftrightarrow{ED}$
12. $\overline{YZ}$ or $\overline{ZY}$

Systematic Review 27D

1. length; width
2. length; width
3. one
4. two
5. $\overrightarrow{EF}$
6. $3+5+5+5+6+7+7+9+16 = 63$
   mean $= 63 \div 9 = 7$
7. median $= 6$
8. mode $= 5$
9. $\frac{15}{100} = \frac{3}{20}$
10. $\frac{375}{1000} = \frac{3}{8}$
11. $0.6 + 0.045 = 0.645$
12. $113 + 0.19 = 113.19$
13. $2.9 + 0.11 = 3.01$
14. 76
15. 55
16. $\frac{1}{12}$
17. $\frac{1}{8}$
18. $10 \div 2 = 5$; $5 \times 4 = 20$;
    $20 \div 2 = 10$; $10 + 7 = 17$

Systematic Review 27E

1. line; ray
2. point
3. line segment
4. endpoint
5. $\overleftrightarrow{GK}$ or $\overleftrightarrow{KG}$
6. $1+1+1+7+7+8+9+14 = 48$
   mean $= 48 \div 8 = 6$
7. median $= 7$
8. mode $= 1$
9. $\frac{65}{100} = \frac{13}{20}$
10. $\frac{1}{1000}$
11. $11.4 - 0.6 = 10.8$
12. $35 - 0.42 = 34.58$
13. $5.15 - 0.9 = 4.25$
14. 890
15. north
16. $\frac{15}{150} = \frac{1}{10}$
17. $3.14 \times 12{,}732 \text{ km} \approx 39{,}978 \text{ km}$
18. $4 \times 2 = 8$; $8 + 1 = 9$;
    $9 - 7 = 2$; $2 \div 2 = 1$

Systematic Review 27F

1. infinite
2. point
3. endpoints
4. endpoint or origin
5. $\overline{MN}$ or $\overline{NM}$
6. $13+13+13+13+14+18 = 84$
   mean $= 84 \div 6 = 14$
7. median $= 13$
8. mode $= 13$
9. $\frac{82}{100} = \frac{41}{50}$
10. $\frac{5}{1000} = \frac{1}{200}$
11. $0.5 \times 0.7 = 0.35$
12. $0.02 \times 0.4 = 0.008$
13. $1.31 \times 0.28 = 0.3668$
14. because it goes the whole way around the city
15. $\frac{1}{7}$
16. $4.5 \text{ m} \times 1.9 \text{ m} \times 6 \text{ m} = 51.3 \text{ m}^3$
17. $7 \text{ ft} \times 1.12 = 7.84 \text{ ft}$
18. $\frac{1}{3} \times 12 = 4$; $4 + 5 = 9$;
    $9 + 1 = 10$; $10 \times 4 = 40$

Lesson Practice 28A

1. length; width
2. two
3. plane
4. congruent
5. equal
6. similar
7. b
8. a
9. d

10. c
11. g
12. e
13. h
14. f

Lesson Practice 28B
1. equal
2. similar
3. congruent
4. plane
5. point
6. geometry
7. d
8. c
9. b
10. a
11. h
12. g
13. e
14. f

Lesson Practice 28C
1. point
2. line, line segment, ray
3. plane
4. equal
5. congruent
6. similar
7. c
8. d
9. a
10. b
11. ▱
12. =
13. ~
14. ≅

Systematic Review 28D
1. congruent
2. length; width
3. similar
4. length; width
5. one
6. length; width
7. two
8. $0.7T + 1.4 = 4.9$
   $0.7T = 4.9 - 1.4$
   $0.7T = 3.5$
   $T = 3.5 \div 0.7$
   $T = 5$
9. $0.7(5) + 1.4 = 4.9$
   $3.5 + 1.4 = 4.9$
   $4.9 = 4.9$
10. $33\frac{1}{3}\%$
11. 50%
12. 75%
13. $0.15 \div 0.9 = 0.17$
14. $1.28 \div 6.2 = 0.21$
15. $0.105 \div 0.5 = 0.21$
16. $3 + 5 + 6 + 6 + 10 = 30$
    $30 \div 5 = 6$
17. 90
18. $6 \div 3 = 2$; $2 \times 4 = 8$;
    $8 - 3 = 5$; $5 + 6 = 11$

Systematic Review 28E
1. h
2. e
3. b
4. f
5. d
6. c
7. a
8. i
9. g
10. $66\frac{2}{3}\%$
11. 25%
12. 20%

13. $0.9133 = 0.91\overline{3}$
$$3\overline{)2.7400}$$
   27000
   400
   300
   100
   900
   10
   9
   1

14. $90.6\overline{6} = 90.\overline{6}$
$$9\overline{)816.00}$$
   81000
   600
   540
   60
   54
   6

15. $4.1\overline{45} = 4.1\overline{45}$
$$11\overline{)45.600}$$
   44000
   1600
   1100
   500
   440
   60
   55
   5

16. 251
17. 176
18. $9 \div 3 = 3$; $3 + 2 = 5$;
    $5 \times 7 = 35$; $35 - 5 = 30$

Systematic Review 28F

1. ∞
2. −
3. →
4. ↔
5. •
6. =
7. ~
8. ≅
9. ▱

10. 40%
11. $33\frac{1}{3}\%$
12. 80%
13. $0.73\frac{13}{19}$
14. $0.85\frac{5}{7}$
15. $0.55\frac{5}{9}$
16. 12 in
17. north; 435
18. $\frac{1}{4} \times 12 = 3$; $3 + 7 = 10$;
    $10 \div 2 = 5$; $5 \times 5 = 25$

Lesson Practice 29A

1. angles
2. rays
3. 90
4. 360
5. vertex
6. circle or turn
7. Greek
8. done
9. ∠QXT; ∠TXQ
10. 90°
11. four

Lesson Practice 29B

1. angle
2. right
3. endpoint
4. circle or turn
5. right angle
6. lines
7. vertex
8. ∠ZYX; ∠XYZ
9. ∠LMN; ∠NML
10. right
11. 90°

Lesson Practice 29C
1. 90
2. 360
3. vertex
4. Greek
5. $\dfrac{1}{4}$
6. 4
7. rays
8. ∠CBA; ∠ABC
9. ∠YRD; ∠DRY
10. 90°
11. 90°

Systematic Review 29D
1. circle
2. 360°
3. 90°
4. similar
5. length; width
6. two
7. congruent
8. $0.85M = 0.425$
   $M = 0.425 \div 0.85$
   $M = 0.5$
9. $0.85(0.5) = 0.425$
   $0.425 = 0.425$
10. $\dfrac{95}{100} = \dfrac{19}{20}$
11. $\dfrac{7}{10}$
12. $\dfrac{685}{1000} = \dfrac{137}{200}$
13. $32 \div 8 = 4$
    $4 \times 3 = 12$
14. $12 \div 6 = 2$
    $2 \times 1 = 2$
15. $72 \div 9 = 8$
    $8 \times 5 = 40$
16. $\dfrac{950 - 2}{950} = \dfrac{948}{950} = \dfrac{474}{475}$
17. $8.5 \text{ lb} + 9.5 \text{ lb} + 10 \text{ lb} + 10 \text{ lb} + 12 \text{ lb} = 50 \text{ lb}$
    $50 \text{ lb} \div 5 = 10 \text{ lb}$
18. $\dfrac{1}{5} \times 15 = 3;\ 3 + 8 = 11;$
    $11 - 2 = 9;\ 9 \times 9 = 81$

Systematic Review 29E
1. b
2. c
3. f
4. e
5. a
6. g
7. d
8. $0.09Q = 0.315$
   $Q = 0.315 \div 0.09 = 3.5$
9. $0.09(3.5) = 0.315$
   $0.315 = 0.315$
10. $\dfrac{24}{100} = \dfrac{6}{25}$
11. $\dfrac{9}{10}$
12. $\dfrac{278}{1000} = \dfrac{139}{500}$
13. $6\dfrac{1}{6} + 8\dfrac{2}{5} = 6\dfrac{5}{30} + 8\dfrac{12}{30} = 14\dfrac{17}{30}$
14. $3 - 2\dfrac{1}{3} = 2\dfrac{3}{3} - 2\dfrac{1}{3} = \dfrac{2}{3}$
15. $7\dfrac{2}{9} + 9\dfrac{4}{5} = 7\dfrac{10}{45} + 9\dfrac{36}{45} =$
    $16\dfrac{46}{45} = 17\dfrac{1}{45}$
16. $3{,}405 \text{ g} \times \dfrac{1 \text{ kg}}{1{,}000 \text{ g}} = 3.405 \text{ kg}$
17. south
18. $20 \div 10 = 2;\ 2 \times 6 = 12;$
    $12 + 3 = 15;\ 15 - 8 = 7$

Systematic Review 29F
1. right
2. $\dfrac{1}{4}$
3. length; width
4. equal
5. line segment
6. rays
7. ∠BCA; ∠ACB

8. $3.3X + 0.79 = 4.42$
    $3.3X = 4.42 - 0.79$
    $3.3X = 3.63$
    $X = 3.63 \div 3.3$
    $X = 1.1$

9. $3.3(1.1) + 0.79 = 4.42$
    $3.63 + 0.79 = 4.42$
    $4.42 = 4.42$

10. $\frac{75}{100} = \frac{3}{4}$

11. $\frac{9}{100}$

12. $\frac{138}{1000} = \frac{69}{500}$

13. $\frac{\cancel{4}}{9} \times \frac{5}{\cancel{8}_2} = \frac{5}{18}$

14. $\frac{\cancel{2}}{\cancel{3}} \times \frac{\cancel{9}^3}{\cancel{10}_5} = \frac{3}{5}$

15. $\frac{3}{\cancel{24}} \times \frac{\cancel{2}}{5} = \frac{3}{10}$

16. $175.62 - $39.59 = 136.03

17. $2.75 \text{ gal/min} \times 12.25 \text{ min} = 33.6875 \text{ gal}$

18. $8 \div 4 = 2$; $2 \times 5 = 10$;
    $10 - 6 = 4$; $4 + 9 = 13$

Lesson Practice 30A

1. acute
2. obtuse
3. straight
4. b
5. d
6. a
7. c
8. right
9. obtuse
10. acute
11. straight
12. straight

Lesson Practice 30B

1. straight
2. obtuse
3. acute
4. c
5. b
6. d
7. a
8. straight
9. obtuse
10. right
11. acute
12. $360° \div 6 = 60°$; acute

Lesson Practice 30C

1. 90°; 180°
2. 0°; 90°
3. 180°
4. 90°
5. b
6. c
7. a
8.
9.
10.
11.
12. 180°

Systematic Review 30D

1. acute
2. length; width
3. 90°
4. similar
5. congruent
6. two
7. vertex
8. $0.07X + 0.52 = 0.527$
    $0.07X = 0.527 - 0.52$
    $0.07X = 0.007$
    $X = 0.1$

9. $0.07(0.1) + 0.52 = 0.527$
    $0.007 + 0.52 = 0.527$
    $0.527 = 0.527$

10. $5 \div 6 = 0.83$
11. $8 \div 13 = 0.62$
12. $4 \div 5 = 0.8$
13. $\dfrac{1}{3} \div \dfrac{4}{7} = \dfrac{1}{3} \times \dfrac{7}{4} = \dfrac{7}{12}$
14. $\dfrac{3}{10} \div \dfrac{3}{5} = \dfrac{\cancel{3}}{\cancel{10}_2} \times \dfrac{\cancel{5}}{\cancel{3}} = \dfrac{1}{2}$
15. $\dfrac{7}{8} \div \dfrac{1}{2} = \dfrac{7}{\cancel{8}_4} \times \dfrac{\cancel{2}}{1} = \dfrac{7}{4} = 1\dfrac{3}{4}$
16. $3.14(26 \text{ in}) = 81.64 \text{ in}$
17. $\dfrac{200}{200+400} = \dfrac{200}{600} = \dfrac{1}{3}$
18. $30 \div 10 = 3$
    $3 \times 4 = 12$
    $12 + 2 = 14$
    $14 - 7 = 7$

Systematic Review 30E

1. b
2. e
3. f
4. c
5. a
6. d
7. g
8. ←——→
9. ⌐
10. $\dfrac{725}{1000} = \dfrac{29}{40}$
11. $\dfrac{3}{10}$
12. $\dfrac{95}{100} = \dfrac{19}{20}$
13. $65 \text{ L} \times \dfrac{1 \text{ kl}}{1{,}000 \text{ L}} = 0.065 \text{ kl}$
14. $21 \text{ mm} \times \dfrac{1 \text{ m}}{1{,}000 \text{ mm}} \times \dfrac{100 \text{ cm}}{1 \text{ m}} = 2.1 \text{ cm}$
15. $0.04 \text{ g} \times \dfrac{100 \text{ cg}}{1 \text{ g}} = 4 \text{ cg}$

16. round: $3.14(6 \text{ in})^2 = 113.04 \text{ in}^2$
    square: $10.5 \text{ in} \times 10.5 \text{ in} = 110.25 \text{ in}^2$
    $113.04 \text{ in}^2 > 110.25 \text{ in}^2$
    round pizza is bigger
17. $\$5.95 \times 1.03 \approx \$6.13$
    $\$10.00 - \$6.13 = \$3.87$
18. $10 \times 4 = 40$
    $40 \div 5 = 8$
    $8 \div 4 = 2$
    $2 \div 2 = 1$

Systematic Review 30F

1. acute
2. obtuse
3. length; width
4. equal
5. $\dfrac{1}{4}$
6. congruent
7. $\angle XTQ$; $\angle QTX$
8. $3+5+5+5+6+7+7+9+16 = 63$
   mean $= 63 \div 9 = 7$
   median $= 6$
   mode $= 5$
9. $9+13+27+27+30+32+37 = 175$
   mean $= 175 \div 7 = 25$
   median $= 27$
   mode $= 27$
10. 350%
11. 1%
12. 45%
13. $0.35 \times 200 = 70$
14. $0.02 \times 16 = 0.32$
15. $2.50 \times 150 = 375$
16. $\dfrac{5}{695} = \dfrac{1}{139}$
17. $6^2 = 6 \times 6 = 36$
18. $8 + 4 = 12$
    $12 - 6 = 6$
    $6 + 3 = 9$
    $9 \times 2 = 18$

Application and Enrichment Solutions

Application and Enrichment 1G

1. sum
2. factors
3. product
4. quotient
5. G, B
6. F, optional: C, E
7. A, optional: E
8. H, optional: E
9. E, optional: A
10. B, optional: E
11. C, optional: E
12. D, optional: E
13. B, optional: A, E
14. 3X
15. 3G
16. 4Y
17. 11A

Application and Enrichment 2G

1. Each expression has a value of 9.
2. always true
3. always true
4. sometimes false
5. always true
6. sometimes false
7. done
8. $20 \times 6 = 120$
   $4 \times 30 = 120$
9. $6 + 7 = 13$
   $9 + 4 = 13$
10. $2 \div 2 = 1$
    $12 \div 3 = 4$
11. $5 - 3 = 2$
    $10 - 2 = 8$

Application and Enrichment 3G

1. 4
2. 5
3. 8
4. 9
5. 5
6. 1, 2, 3, 4, 6, 8, 12, 24
7. 1, 3, 9
8. 1, 3, 5, 9, 15, 45
9. 1, 2, 4, 8, 16
10. $6(1) + 6(2)$
11. $5(5) + 5(6)$
12. $3(3) + 3(7)$
13. $6(1 + 2) = 6(3) = 18$
    $6 + 12 = 18$
14. $5(5 + 6) = 5(11) = 55$
    $25 + 30 = 55$
15. $3(3 + 7) = 3(10) = 30$
    $9 + 21 = 30$

Application and Enrichment 4G

1. $3(7) = 21$
   $18 + 3 = 21$ The answers agree.
2. $4(2 + 5) = 4(7) = 28$
   $4(2) + 4(5) = 8 + 20 = 28$
3. $2(10 + 6) = 2(16) = 32$
   $2(10) + 2(6) = 20 + 12 = 32$
4. $6(2 + 3) = 6(5) = 30$
   $6(2) + 6(3) = 12 + 18 = 30$
5. $7(8 + 4) = 7(12) = 84$
   $7(8) + 7(4) = 56 + 28 = 84$
6. $4(2) + 4(B) = 8 + 4B$
7. $8(A) + 8(4) = 8A + 32$
8. $2(X) + 2(Z) = 2X + 2Z$
9. $7(A) + 7(2Y) = 7A + 14Y$
10. $2(3) + 2(5B) = 2(3 + 5B)$
11. $5(3) + 5(7X) = 5(3 + 7X)$
12. $9(1) + 9(8Y) = 9(1 + 8Y)$

Application and Enrichment 5G

1. Answers will vary. Both sides of final equation should match. Example:
   $1 + 1 = 2(1); 2 = 2$
   $5 + 5 = 2(5); 10 = 10$
2. Answers will vary. Both sides of final equation should match.
3. Answers will vary. Both sides of final equation should match.

APPLICATION AND ENRICHMENT SOLUTIONS

4. 3(2) + 4 = 4 + 3(2)
   6 + 4 = 4 + 6
   10 = 10; true
5. 5(8 - 6) = 5(8) - 30
   5(2) = 40 - 30
   10 = 10; true
6. 4(3) = 2(3) + 2
   12 = 6 + 2
   12 ≠ 8; false
7. 5(2) ÷ 5 = 5 ÷ 5(2)
   10 ÷ 5 = 5 ÷ 10
   2 ≠ 1/2; false
8. done
9. 3, 6, 9, 12, 15 . . .
   4, 8, 12, 16 . . .
   The LCM of 3 and 4 is 12.
10. 6, 12, 18, 24, 30 . . .
    10, 20, 30, 40 . . .
    The LCM of 6 and 10 is 30.
11. 4, 8, 12, 16 . . .
    12, 24, 36, 48 . . .
    The LCM of 4 and 12 is 12.
12. 9, 18, 27, 36, 45 . . .
    12, 24, 36, 48 . . .
    The LCM of 9 and 12 is 36.

Application and Enrichment 6G

1. 10 years
2. 24 years
3. 14 years
4. 10, 15, 24 years
5-8. On graph, 1 has 1 dot, 2 has 2, 3 has 3, 7 has 1 dot.
5. 3
6. 1, 7
7. 1 + (2 × 2) + (3 × 3) + 7 = 1 + 4 + 9 + 7 = 21 pets
8. 21 ÷ 7 = 3 pets average

Application and Enrichment 7G

1. 3 kg
2. 5 - 3 = 2 kg
3. 8 kg

4. birth to three months
5.

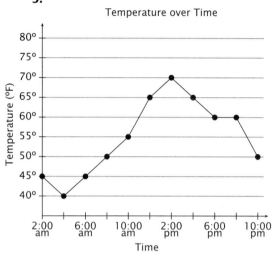

6. 40° at 4:00 AM
7. 70° at 2:00 PM

Application and Enrichment 8G

1. Freddie, 4 flies
2. Billy, 14 flies
3. 4 + 8 + 6 + 14 = 32 flies
4. 32 ÷ 4 = 8 flies
5. no
6.

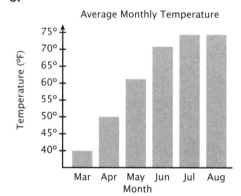

7. March
8. July, August
9. 74° - 40° = 34°

Application and Enrichment 9G

1. 31-40 years

2. 61-70 years
3. Sample answer: Yes; many of the employees are around the age at which people are most likely to have young families (mid 20s to late 30s).
4. Sample answer: No; the fact that employees are the right age to have young families does not necessarily mean they are married or have children.
5. 60°-65°: 6 times
   65°-70°: 3 times
   70°-75°: 9 times
   75°-80°: 1 time
   80°-85°: 1 time
6.

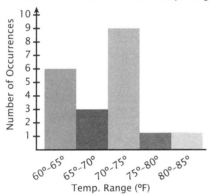

Number of Occurrences over Temp. Range

Application and Enrichment 10G

1. $3.29
2. $16.12
3. $5.70
4. $0.35
5. 25.4 inches
6. 6.92 m
7. 125 mi
8. 0.40 cm
9. 1,568.21 miles
10. 1,600 miles

Application and Enrichment 11G

1.

trunks	legs
1	4
2	8
3	12
4	16

2. count by one/add one
3. skip count by 4/multiply by 4
4. 4

dishes	minutes
3	1
6	2
9	3
12	4

5. 8 minutes; divide by 3 or multiply by 1/3.
6. 8 dishes ÷ 4 min = 2 dishes/min

Application and Enrichment 12G

1. 1/10 × 300 = 30 votes
2. 16 red for every 20 yellow

red	4	8	12	16
yellow	5	10	15	20

3. 8/4 = 8 ÷ 4 = 2 minutes per pan
4. 4 × 15 = 60
   2 × 4 = 8 incorrect

Application and Enrichment 13G

1.

gallons of green	2	4	6	8	10
total gallons	3	6	9	12	15

2. 8 gallons
3. 12 − 8 = 4 gallons

APPLICATION AND ENRICHMENT SOLUTIONS

4.

cranberry	3	6	9	12	15
mixture	5	10	15	20	25

5. 12 cups
6. 20 − 12 = 8 cups

Application and Enrichment 14G

1.

Miles	Hours	MPH
200	4	50
300	6	50
150	3	50
400	8	50

2.

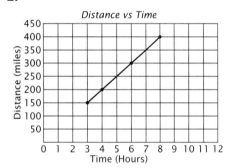

3. done
4. 80 mi/2 hr = 40 mi/hr
40 mi/hr × 3 hr = 120 mi
5. 15 mi/5 hr = 3 mi/hr
3 mi/hr × 3 hr = 9 mi
6. done
7. 1 oz/28 g = ?/56 g
28 g × 2 = 56 g; 1 oz × 2 = 2 oz
8. 1 gal/4 qt = ? gal/32 qt
4 × 8 = 32; 1 gal × 8 = 8 gal
9. (6 ft/1) × (1 yd/3 ft) = 6/3 yd = 2 yd
10. (5 kg/1) × (1000 g/1 kg) = 5,000 g
11. (12 qt/1) × (1 gal/4 qt) = 3 gal
12. (84 g/1) × (1 oz/28 g) = 3 oz
13. (9 km/1) × (1,000 m/1 km) = 9,000 m
14. (55 kg/1) × (2.2 lb/1 kg) = 121 lb

Application and Enrichment 15G

1.

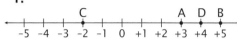

2. <
3. <
4. >
5. <
6. 20
7. 10
8. 15

Application and Enrichment 16G

1. $55
2. −32 > −120; Justin is camping at the higher location.
3. Jerry; the absolute value (distance from zero) of −120 is greater than the absolute value of −32.
4. 31° > −13°
5. |31| > |−13| or |−13| < |31|
6.

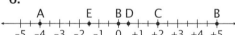

7.

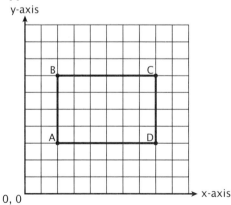

20 units

APPLICATION AND ENRICHMENT SOLUTIONS

8. 4 units
9. 6 units
10. 6 + 6 + 4 + 4 = 20 units
    Adding the lengths of the sides yields the same answer as counting.

Application and Enrichment 17G

1.

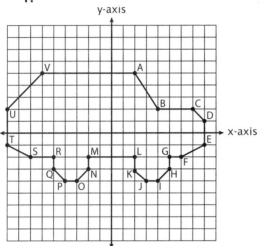

Application and Enrichment 18G

1.

Wheat Flour	Rolled Oats
5	2
10	4
15	6
20	8

2. 5 cups
3. 2 cups
4.

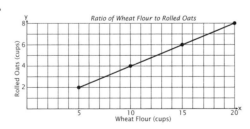

5.

Feet	Minutes
3	1
6	2
9	3
12	4
15	5

6. 2
7. 3
8. 4
9. skip counting by 3
10.

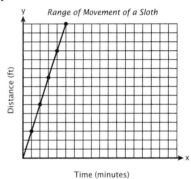

Application and Enrichment 19G

1. $0.8A = 160$
   $A = 160 \div 0.8;\ A = 200\ yd^2$
2. $0.7D = 49$
   $D = 49 \div 0.7;\ D = 70\ in$
3. $0.25(M) = \$15$
   $M = \$15 \div 0.25 = \$60$
4. $0.75P = \$261$
   $P = \$261 \div 0.75 = \$348$
5. $0.3T = 3$
   $T = 3 \div 0.3 = 10\ hours$
6. false, false, true
7. true, true, true
8. $0.5(0.4) = 0.2;\ <$
9. $0.5(0.4) = 2;\ <$
10. $0.5(40) = 20;\ >$

Application and Enrichment 20G

1. $2\pi r \approx 2(3.14)(2) \approx 12.56$ in
   $\pi(2^2) \approx 12.56$ in^2
2. $2\pi r \approx 2(3.14)(4) \approx 25.12$ in
   $\pi(4^2) \approx 3.14(16) \approx 50.24$ in^2
3. 2
4. 4
5. $2\pi r \approx 2(3.14)(3) \approx 18.84$ in
   $\pi(3^2) \approx 3.14(9) \approx 28.26$ in^2
6. $2\pi r \approx 2(3.14)(6) \approx 37.68$ in
   $\pi(6^2) \approx 3.14(36) \approx 113.04$ in^2
7. 2
8. 4

Application and Enrichment 21G

1. (4, 0)
2. (4, 5)
3. (0, 5)
4. (0, 0)
5.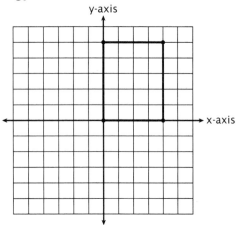
6. 4
7. 5
8. $4 \times 5 = 20$

9.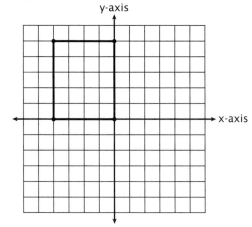
10. −4
11. 4 blocks
12. 5
13. 5 blocks
14. $4 \times 5 = 20$ square blocks

Application and Enrichment 22G

1. $5C = 20$
2. $C = 4$ cars
3. $\$3.50(D) = M$
   $\$3.50(5) = \$17.50$
4. $\$3.50(12) = \$42$
5. $\$3.50(D) = \$35$
   $D = 10$ days and sandwiches
6.

Radius	1 cm	2 cm	4 cm	8 cm
C = 2(3.14)r	6.28 cm	12.56 cm	25.12 cm	50.24 cm

7.

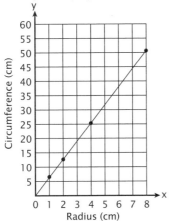

Application and Enrichment 23G

1. 3 ft × 4 ft × 5 ft = 60 ft³
2. bottom = 15 ft²
   top = 15 ft²
   front = 20 ft²
   back = 20 ft²
   side 1 = 12 ft²
   side 2 = 12 ft²
3. surface area = 94 ft²
4. 96 ft² + 120 ft² + 96 ft² + 120 ft² + 80 ft² + 80 ft² = 592 ft²
5. 600 ft² × (1 can/100 ft²) = 6 cans
6. triangles:
   (1/2)(6 in × 4 in) = 12 in²
   square: 6 in × 6 in = 36 in²
   (12 in² × 4) + 36 in² = 84 in²

Application and Enrichment 24G

1. 6
2. 12
3. yes
4. 6 + 6 + 6 + 6 + 4 + 4 = 32 squares
5. 1.5 in × 1.5 in = 2.25 in²

6.

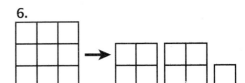

7. (1.5 in)(1.5 in)(1 in) = 2.25 in³
   Since one unit block is about half an inch on a side, it takes about eight to make a cubic inch. You should have used 18 blocks to build the shape, since 18 ÷ 8 = 2.25.
8. (1.5 in)(1.5 in)(1.5 in) = 3.375 in³
   27 blocks = 3 groups of eight blocks (3 in³) with 3 left over (0.375 in³)
9. T < 100° C

10.

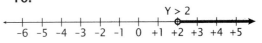

11.

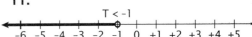

Application and Enrichment 25G

1. lowest = 2
   highest = 10
   spread = 10 − 2 = 8
2. 2, 3, 4, 5, 5, 6, 6, 7, 7, 7, 8, 8, 8, 8, 9, 9, 10
   median = 7
3. mode = 8
4. asymmetrical

5.

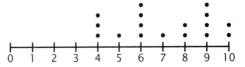

6. 10 − 4 = 6
7. 4, 4, 4, 5, 6, 6, 6, 6, 7, 8, 8, 9, 9, 9, 9, 10, 10
   median = 7

8. 2 peaks
9. 120 ÷ 17 = 7.1

Application and Enrichment 26G

1. 4 + 2 + 4 + 6 + 2 + 1 = 19 people
2. 2 + 4 + 2 + 4 + 1 + 1 = 14 people
3. 19 + 14 = 33 people
4. no
5. B
6. B
7. Sample answer: "Do you support spending town funds for the improvement of Main Street?"
8. inches
9. feet
10. Answers will vary.

Application and Enrichment 27G

1. (1 + 3 + 3 + 5 + 6 + 9) ÷ 6 = 27 ÷ 6 = 4.5 average per family
2.

J	L	B	R	T	M
1	3	3	5	6	9

3.

J	L	B	R	T	M
1	3	3	5	6	9
3.5	1.5	1.5	0.5	1.5	4.5

4. (3.5 + 1.5 + 1.5 + 0.5 + 1.5 + 4.5) ÷ 6 = 13 ÷ 6 = 2 1/6 ≈ 2.17

Application and Enrichment 28G

1. 2, 3, 3, 3, 3, 4, 4, 4, 5, 5, 6, 7, 14; median = 4
2. 2, 3, 3, 3, 3, 4
   1st quartile = (3 + 3) ÷ 2 = 3
3. 4, 5, 5, 6, 7, 14
   3rd quartile = (5 + 6) ÷ 2 = 5.5
4. IQR = 5.5 − 3 = 2.5

5. Range = 14 − 2 = 12 much more than twice the size of the IQR even though it is only twice as many data.
6. Mean = (2 + 3 + 3 + 3 + 3 + 4 + 4 + 4 + 5 + 5 + 6 + 7 + 14) ÷ 13 = 4.8, which is 0.8 longer than the median

Application and Enrichment 29G

1.

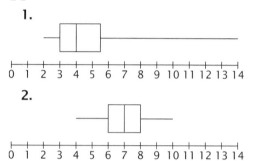

2.

3. symmetrical

Application and Enrichment 30G

The student should review the main ideas and use them when observing and recording real-life data.

Test Solutions

Lesson Test 1

1. $1 \times 1 \times 1 \times 1 = 1$
2. $10 \times 10 = 100$
3. $13 \times 13 = 169$
4. $3 \times 3 \times 3 = 27$
5. $12 \times 12 = 144$
6. $7 \times 7 = 49$
7. $5^2 = 25$
8. $4^3 = 64$
9. $2^4 = 16$
10. $6 \times 6 \times 6 \times 6 = 6^4$
11. $15 \times 15 = 15^2$
12. $9 \times 9 \times 9 = 9^3$
13. $8 \times 8 \times 8 = 8^3$ or $512 = 8^3$
14. $512 = 8^3$ or $8 \times 8 \times 8 = 8^3$
15. $96 \div 4 = 24$
    $24 \times 1 = 24$
16. $88 \div 8 = 11$
    $11 \times 3 = 33$
17. $100 \div 10 = 10$
    $10 \times 7 = 70$
18. $60 \text{ min} \div 4 = 15 \text{ min}$
    $15 \text{ min} \times 3 = 45 \text{ min}$
19. $24 \div 8 = 3$
    $3 \times 1 = 3$ apples
20. $20 \div 10 = 2$
    $2 \times 9 = 18$ people

Lesson Test 2

1. $10 \times 10 \times 10 \times 10 = 10,000$
2. $10 \times 10 = 100$
3. $10 \times 10 \times 10 = 1,000$
4. $10 = 10^1$
5. $1,000,000 = 10^6$
6. $1 = 10^0$
7. $1 \times 100 + 2 \times 10 + 5 \times 1$
8. $3 \times 1,000 + 9 \times 100 + 3 \times 10 + 2 \times 1$
9. $9 \times 10^3 + 2 \times 10^1 + 8 \times 10^0$
10. $1 \times 10^4 + 6 \times 10^3 + 9 \times 10^1$
11. $83,200$
12. $2,680$
13. $8 \times 8 = 64$
14. $2 \times 2 \times 2 = 8$
15. $5^3 = 125$
16. $\dfrac{1}{8} = \dfrac{2}{16} = \dfrac{3}{24} = \dfrac{4}{32}$
17. $\dfrac{4}{5} = \dfrac{8}{10} = \dfrac{12}{15} = \dfrac{16}{20}$
18. $24 \text{ hr} \div 6 = 4 \text{ hr}$
    $4 \text{ hr} \times 3 = 12 \text{ hr}$
19. $24 \text{ hr} - 12 \text{ hr} = 12 \text{ hr}$
20. $1/2$ of $60 \text{ min} = 30 \text{ min}$
    $1/3$ of $60 \text{ min} = 20 \text{ min}$
    $1/4$ of $60 \text{ min} = 15 \text{ min}$
    $1/4$ of $60 \text{ min} = 15 \text{ min}$
    $1/2$ of $60 \text{ min} = 30 \text{ min}$
    $30 + 20 + 15 + 15 + 30 = 110 \text{ min}$

Lesson Test 3

1. $3 \times 10^1 + 4 \times 10^0 + 6 \times \dfrac{1}{10^1}$
2. $1 \times \dfrac{1}{10^1} + 2 \times \dfrac{1}{10^2} + 4 \times \dfrac{1}{10^3}$
3. $5,417.04$
4. $1,000.875$
5. dimes; cents;
   $3.00 + 0.0 + 0.05 = 3.05$
6. dollars; dimes; cent;
   $6.00 + 0.4 + 0.01 = 6.41$
7. $10 \times 10 \times 10 = 1,000$
8. $1 \times 1 \times 1 \times 1 \times 1 \times 1 \times 1 = 1$
9. $6 \times 6 = 36$
10. $11 \times 11 = 121$
11. $\dfrac{2}{9} = \dfrac{4}{18} = \dfrac{6}{27} = \dfrac{8}{36}$
12. $\dfrac{1}{7} = \dfrac{2}{14} = \dfrac{3}{21} = \dfrac{4}{28}$
13. $\dfrac{3}{12} = \dfrac{1}{4}$
14. $\dfrac{2}{50} = \dfrac{1}{25}$
15. $\dfrac{8}{64} = \dfrac{1}{8}$

16. $\dfrac{15}{90} = \dfrac{1}{6}$
17. $4.63
18. 100 cents ÷ 10 = 10 cents
    10 cents × 5 = 50 cents
19. $\dfrac{50}{100} = \dfrac{1}{2}$
20. $5 \times 5 = 5^2$ blocks

Lesson Test 4

1. $\overset{1}{}6.7$
    $+\ 5.4$
    $\overline{12.1}$
2. 2.0
    $+\ \ .2$
    $\overline{2.2}$
3. 6.24
    $+\ 8.40$
    $\overline{14.64}$
4. $\overset{1}{}5.28$
    $+2.05$
    $\overline{7.33}$
5. $1 \times 1 \times 1 = 1$
6. $10 \times 10 = 100$
7. $6 \times 6 \times 6 = 216$
8. $7 \times 7 = 49$
9. 8,400.2
10. 269.005
11. $\dfrac{4}{5} = \dfrac{8}{10} = \dfrac{12}{15} = \dfrac{16}{20}$
12. $\dfrac{5}{9} = \dfrac{10}{18} = \dfrac{15}{27} = \dfrac{20}{36}$
13. $\dfrac{1}{6} + \dfrac{2}{9} = \dfrac{9}{54} + \dfrac{12}{54} = \dfrac{21}{54} = \dfrac{7}{18}$
14. $\dfrac{1}{5} + \dfrac{7}{10} = \dfrac{10}{50} + \dfrac{35}{50} = \dfrac{45}{50} = \dfrac{9}{10}$
15. $\dfrac{1}{3} + \dfrac{3}{8} = \dfrac{8}{24} + \dfrac{9}{24} = \dfrac{17}{24}$
16. $4.75 + $6.29 = $11.04
    $11.04 > $11.00; yes
17. $1.2 + .75 + 1.15 = 3.1$ carats
18. $\dfrac{4}{10} = \dfrac{2}{5}$

19. $\dfrac{5}{8} + \dfrac{1}{6} = \dfrac{30}{48} + \dfrac{8}{48} = \dfrac{38}{48} = \dfrac{19}{24}$
20. 30 days ÷ 5 = 6 days
    6 days × 3 = 18 days

Lesson Test 5

1. $^{6}\!\!\cancel{7}.^{1}25$
    $-\ 2.\ 8$
    $\overline{4.\ 45}$
2. $^{4}\!\!\cancel{5}.^{1}07$
    $-\ 2.\ 11$
    $\overline{2.\ 96}$
3. $^{5}\!\!\cancel{6}.^{1}\!\!\cancel{2}^{1}49$
    $-\ 5.\ 371$
    $\overline{0.\ 878}$
4. $\overset{1}{}8.2$
    $+0.9$
    $\overline{9.1}$
5. $\overset{1}{}6.14$
    $+0.395$
    $\overline{6.535}$
6. $\overset{1}{}14.9$
    $+\ 0.124$
    $\overline{15.024}$
7. $\dfrac{2}{5} + \dfrac{1}{8} = \dfrac{16}{40} + \dfrac{5}{40} = \dfrac{21}{40}$
8. $\dfrac{3}{7} + \dfrac{2}{9} = \dfrac{27}{63} + \dfrac{14}{63} = \dfrac{41}{63}$
9. $\dfrac{2}{3} + \dfrac{1}{6} = \dfrac{12}{18} + \dfrac{3}{18} = \dfrac{15}{18} = \dfrac{5}{6}$
10. 29,000.1
11. 580.034
12. $\dfrac{3}{4} - \dfrac{1}{5} = \dfrac{15}{20} - \dfrac{4}{20} = \dfrac{11}{20}$
13. $\dfrac{7}{8} - \dfrac{2}{3} = \dfrac{21}{24} - \dfrac{16}{24} = \dfrac{5}{24}$
14. $\dfrac{1}{4} - \dfrac{1}{6} = \dfrac{6}{24} - \dfrac{4}{24} = \dfrac{2}{24} = \dfrac{1}{12}$
15. $20.00 - $15.24 = $4.76
16. $7.35 + $2.35 + $1.17 + $1.00 = $11.87
17. $35.00 - $11.87 = $23.13
18. 3.2 hr + 2.3 hr + 1.05 hr = 6.55 hr

19. 100 cents ÷ 10 = 10 cents
    10 cents × 2 = 20 cents
20. $\frac{3}{4} - \frac{1}{10} = \frac{30}{40} - \frac{4}{40} = \frac{26}{40} = \frac{13}{20}$

18. 1,000 g
19. 6.1 min − 5.7 min = 0.4 min
20. $3 \times 10^3 + 4 \times 10^2 + 1 \times \frac{1}{10^1} + 2 \times \frac{1}{10^2}$

Lesson Test 6

1. $\frac{1 \text{ kilogram(kg)}}{1{,}000 \text{ grams(g)}}$ $\frac{1 \text{ hectogram(hg)}}{100 \text{ grams(g)}}$
   $\frac{1 \text{ dekagram(dag)}}{10 \text{ grams(g)}}$ $\frac{1 \text{ gram(g)}}{1 \text{ gram(g)}}$
2. $\frac{1 \text{ kiloliter(kl)}}{1{,}000 \text{ liters(L)}}$ $\frac{1 \text{ hectoliter(hl)}}{100 \text{ liters(L)}}$
   $\frac{1 \text{ dekaliter(dal)}}{10 \text{ liters(L)}}$ $\frac{1 \text{ liter(L)}}{1 \text{ liter(L)}}$
3. $\frac{1 \text{ kilometer(km)}}{1{,}000 \text{ meters(m)}}$ $\frac{1 \text{ hectometer(hm)}}{100 \text{ meters(m)}}$
   $\frac{1 \text{ dekameter(dam)}}{10 \text{ meters(m)}}$ $\frac{1 \text{ meter(m)}}{1 \text{ meter(m)}}$
4. c
5. a
6. b
7. 1.2
   − 0.4
   0.8
8. $^4\cancel{5}.^10$
   − 2.8
   2.2
9. 25.32
   + 1.06
   26.38
10. 7.$^3\cancel{4}^1$2
    − 2. 3 9
    5. 0 3
11. $\frac{1}{5} + \frac{1}{9} = \frac{9}{45} + \frac{5}{45} = \frac{14}{45}$
12. $\frac{3}{8} - \frac{1}{4} = \frac{12}{32} - \frac{8}{32} = \frac{4}{32} = \frac{1}{8}$
13. $\frac{7}{9} - \frac{2}{7} = \frac{49}{63} - \frac{18}{63} = \frac{31}{63}$
14. $5\frac{2}{3} = \frac{15}{3} + \frac{2}{3} = \frac{17}{3}$
15. $2\frac{6}{7} = \frac{14}{7} + \frac{6}{7} = \frac{20}{7}$
16. $1\frac{5}{8} = \frac{8}{8} + \frac{5}{8} = \frac{13}{8}$
17. kilometers

Lesson Test 7

1. $\frac{1 \text{ gram(g)}}{1 \text{ gram(g)}}$ $\frac{1 \text{ decigram(dg)}}{\frac{1}{10} \text{ gram(g)}}$
   $\frac{1 \text{ centigram(cg)}}{\frac{1}{100} \text{ gram(g)}}$ $\frac{1 \text{ milligram(mg)}}{\frac{1}{1{,}000} \text{ gram(g)}}$
2. $\frac{1 \text{ liter(L)}}{1 \text{ liter(L)}}$ $\frac{1 \text{ deciliter(dl)}}{\frac{1}{10} \text{ liter(L)}}$
   $\frac{1 \text{ centiliter(cl)}}{\frac{1}{100} \text{ liter(L)}}$ $\frac{1 \text{ milliliter(ml)}}{\frac{1}{1{,}000} \text{ liter(L)}}$
3. $\frac{1 \text{ meter(m)}}{1 \text{ meter(m)}}$ $\frac{1 \text{ decimeter(dm)}}{\frac{1}{10} \text{ meter(m)}}$
   $\frac{1 \text{ centimeter(cm)}}{\frac{1}{100} \text{ meter(m)}}$ $\frac{1 \text{ millimeter(mm)}}{\frac{1}{1{,}000} \text{ meter(m)}}$
4. kilo
5. hecto
6. deka
7. 40 ÷ 2 = 20
   20 × 1 = 20
8. 64 ÷ 8 = 8
   8 × 3 = 24
9. 14 ÷ 7 = 2
   2 × 2 = 4
10. $\frac{17}{8} = \frac{16}{8} + \frac{1}{8} = 2\frac{1}{8}$
11. $\frac{35}{4} = \frac{32}{4} + \frac{3}{4} = 8\frac{3}{4}$
12. $\frac{29}{3} = \frac{27}{3} + \frac{2}{3} = 9\frac{2}{3}$
13. Pippin
14. 5 quarts
15. kilograms
16. kiloliter
17. half a mile
18. two pounds
19. 1.2 in + 2.14 in + 1.75 in = 5.09 in
20. 50.9 in − 5.09 in = 45.81 in

Lesson Test 8

1. 45,000 liters
2. 6,000 mg
3. 1,300 mm
4. b
5. a
6. c
7. e
8. f
9. d
10. c
11. a
12. b
13. $6\frac{3}{4} + 3\frac{1}{5} = 6\frac{15}{20} + 3\frac{4}{20} = 9\frac{19}{20}$
14. $11\frac{1}{9} + 4\frac{2}{7} = 11\frac{7}{63} + 4\frac{18}{63} = 15\frac{25}{63}$
15. $21\frac{2}{5} + 7\frac{5}{6} = 21\frac{12}{30} + 7\frac{25}{30} = 28\frac{37}{30} = 29\frac{7}{30}$
16. They are the same.
17. 1 kilogram
18. 5 grams
19. 500 centigrams
20. $4\frac{1}{2} + 4\frac{1}{2} = 8\frac{2}{2} = 9$ pies

Unit Test I

1. $1 \times 1 \times 1 \times 1 \times 1 \times 1 = 1$
2. $10 \times 10 \times 10 = 1,000$
3. $9 \times 9 = 81$
4. $3 \times 10 + 5 \times 1 + 2 \times \frac{1}{10} + 4 \times \frac{1}{100}$
5. $9 \times 10^4 + 1 \times \frac{1}{10^1} + 6 \times \frac{1}{10^2}$
6. 4,168.321
7. ${}^5\!6.^13\,9$
   $-3.5\,4$
   $2.8\,5$
8. ${}^4\!5.^9\!0^1\!0$
   $-2.1\,5$
   $2.8\,5$
9. $1.^7\!8^9\!0^1\!1$
   $-0.9\,9\,9$
   $0.8\,0\,2$
10. 8.3
    $+0.4$
    8.7
11. 1.23
    $+0.147$
    1.377
12. 91.2
    $+0.608$
    91.808
13. $96 \div 8 = 12$
    $12 \times 7 = 84$
14. $81 \div 9 = 9$
    $9 \times 5 = 45$
15. $\frac{6}{11} = \frac{12}{22} = \frac{18}{33} = \frac{24}{44}$
16. $\frac{9}{10} = \frac{18}{20} = \frac{27}{30} = \frac{36}{40}$
17. $\frac{1}{6} - \frac{1}{7} = \frac{7}{42} - \frac{6}{42} = \frac{1}{42}$
18. $\frac{3}{4} - \frac{2}{5} = \frac{15}{20} - \frac{8}{20} = \frac{7}{20}$
19. $\frac{7}{9} - \frac{4}{9} = \frac{3}{9} = \frac{1}{3}$
20. $1\frac{2}{3} + 4\frac{1}{6} = 1\frac{12}{18} + 4\frac{3}{18} = 5\frac{15}{18} = 5\frac{5}{6}$
21. $10\frac{3}{9} + 7\frac{1}{4} = 10\frac{12}{36} + 7\frac{9}{36} = 17\frac{21}{36} = 17\frac{7}{12}$
22. $34\frac{5}{6} + 6\frac{1}{8} = 34\frac{40}{48} + 6\frac{6}{48} = 40\frac{46}{48} = 40\frac{23}{24}$
23. 1,200 cm
24. 900,000 cg
25. 220 ml
26. gram
27. kilogram
28. meter
29. 52.5 mi + 15.06 mi = 67.56 mi
30. $50.00 − $16.95 = $33.05

Lesson Test 9

1. $\begin{array}{r} 1.3 \\ \times\ 0.2 \\ \hline 0.26 \end{array}$

2. $\begin{array}{r} 2.3 \\ \times 1.2 \\ \hline 4\ 6 \\ 2\ 3\ \\ \hline 2.7\ 6 \end{array}$

3. $\begin{array}{r} 0.7 \\ \times\ 0.1 \\ \hline 0.07 \end{array}$

4. $\begin{array}{r} 1.1 \\ \times\ 0.6 \\ \hline 0.66 \end{array}$

5. $\begin{array}{r} 1.4 \\ \times 1.2 \\ \hline 2\ 8 \\ 1\ 4\ \\ \hline 1.6\ 8 \end{array}$

6. $\begin{array}{r} 2.0 \\ \times\ 0.4 \\ \hline 0.80 \end{array}$

7. 250 hg
8. 800 cm
9. 310 ml
10. liter
11. kilometer
12. kilogram
13. $12\frac{1}{4} - 3\frac{1}{5} = 12\frac{5}{20} - 3\frac{4}{20} = 9\frac{1}{20}$
14. $9 - 2\frac{7}{8} = 8\frac{8}{8} - 2\frac{7}{8} = 6\frac{1}{8}$
15. $21\frac{1}{3} - 6\frac{3}{4} = 21\frac{4}{12} - 6\frac{9}{12} =$
    $20\frac{16}{12} - 6\frac{9}{12} = 14\frac{7}{12}$
16. 10 grams
17. 4.2 mi × 0.2 = 0.84 mi
18. 3.3 lb + 2.75 lb + 4.09 lb = 10.14 lb
19. $4 - 1\frac{5}{6} = 3\frac{6}{6} - 1\frac{5}{6} = 2\frac{1}{6}$ pies
20. 25.6 mi − 19.8 mi = 5.8 mi

Lesson Test 10

1. $\begin{array}{r} 6.24 \\ \times 0.4 \\ \hline 2.496 \end{array}$

2. $\begin{array}{r} 4.4 \\ \times 1.7 \\ \hline 3.08 \\ 4.4\ \\ \hline 7.48 \end{array}$

3. $\begin{array}{r} 0.67 \\ \times 0.12 \\ \hline 0.0134 \\ 0.067\ \\ \hline 0.0804 \end{array}$

4. $\begin{array}{r} 7.18 \\ \times 0.7 \\ \hline 5.026 \end{array}$

5. $\begin{array}{r} 3.3 \\ \times 1.5 \\ \hline 1.65 \\ 3.3\ \\ \hline 4.95 \end{array}$

6. $\begin{array}{r} 0.85 \\ \times 0.01 \\ \hline 0.0085 \end{array}$

7. 900,000 cg
8. 48,000 mm
9. 12.17 + 147.09 = 159.26
10. 5.13 + 0.26 = 5.39
11. 18.7 − 4.8 = 13.9
12. $4\frac{3}{4} + 1\frac{1}{10} = 4\frac{30}{40} + 1\frac{4}{40} =$
    $5\frac{34}{40} = 5\frac{17}{20}$
13. $13\frac{1}{2} - 4\frac{1}{8} = 13\frac{8}{16} - 4\frac{2}{16} =$
    $9\frac{6}{16} = 9\frac{3}{8}$
14. $20 - 10\frac{1}{5} = 19\frac{5}{5} - 10\frac{1}{5} = 9\frac{4}{5}$
15. $\frac{5}{6} \times \frac{1}{4} = \frac{5}{24}$
16. $\frac{3}{\cancel{8}_4} \times \frac{\cancel{2}}{3} = \frac{1}{4}$
17. $\frac{5}{\cancel{6}_3} \times \frac{\cancel{2}}{3} = \frac{5}{9}$

18. $4 \times \$9.24 = \$36.96$
19. $642 \text{ km} \times 0.6 \text{ mi/km} = 385.2 \text{ mi}$
20. $\$59.50 \times 1.75 \approx \$104.13$

Lesson Test 11

1. 0.35
2. 0.06
3. 0.19
4. 0.15
5. 0.58
6. 0.07
7. $\frac{2}{100} = \frac{1}{50}$
8. $\frac{23}{100}$
9. $\frac{44}{100} = \frac{11}{25}$
10. $1 \div 5 = 0.20 = 20\%$
11. $1 \div 2 = 0.50 = 50\%$
12. $1 \div 4 = 0.25 = 25\%$
13. $3 \div 4 = 0.75 = 75\%$
14.
```
    9.30
  × 0.01
    9 3 0
    0 0 0
  0.0 9 3 0
```
15.
```
    3.5
  × 4.5
    1 5 5
    1 2 0
   15.7 5
```
16.
```
    2.56
  × 0.81
    2 5 6
   60 8
  2.07 36
```
17. $\frac{\cancel{30}^{15}}{7} \times \frac{3}{\cancel{2}} = \frac{45}{7} = 6\frac{3}{7}$
18. $\frac{\cancel{5}}{3} \times \frac{34}{\cancel{5}} = \frac{34}{3} = 11\frac{1}{3}$
19. $\$15.96 \times 0.15 = \$2.394$
    $\$15.96 + 2.39 = \$18.35$
20. $20 \text{ km} \times 0.60 \text{ mi/km} = 12 \text{ mi}$

Lesson Test 12

1. $\frac{300}{100} + \frac{75}{100} = \frac{375}{100} = 375\% = 3.75$
2. $\frac{100}{100} + \frac{40}{100} = \frac{140}{100} = 140\% = 1.40$
3. $\frac{600}{100} + \frac{90}{100} = \frac{690}{100} = 690\% = 6.90$
4. $\frac{200}{100} + \frac{50}{100} = \frac{250}{100} = 250\% = 2.50$
5. $0.35 = \frac{35}{100} = \frac{7}{20}$
6. $0.11 = \frac{11}{100}$
7. yard
8. ounce
9. inch
10. inch
11. pounds
12. mile
13. quarts
14. $\frac{3}{5} \div \frac{1}{3} = \frac{9}{15} \div \frac{5}{15} = \frac{9 \div 5}{1} = \frac{9}{5} = 1\frac{4}{5}$
15. $\frac{3}{4} \div \frac{7}{8} = \frac{24}{32} \div \frac{28}{32} = \frac{24 \div 28}{1} = \frac{24}{28} = \frac{6}{7}$
16. $\frac{2}{5} \div \frac{3}{8} = \frac{16}{40} \div \frac{15}{40} = \frac{16 \div 15}{1} = \frac{16}{15} = 1\frac{1}{15}$
17. $\$11.25 \times 1.10 = \$12.38$ (rounded)
18. $25 \times 4.00 = 100$ cards
19. $\frac{1}{2} \div \frac{1}{10} = \frac{10}{20} \div \frac{2}{20} = \frac{10 \div 2}{1} = 5$ guests
20. $\$75.00 \times 0.25 = \$18.75$

Lesson Test 13

1. ■ orchestra
    ▨ chorus
    □ both

2. $2 \text{ hr} \times 60 \text{ min/hr} = 120 \text{ min}$
    $120 \text{ min} \times 0.05 = 6 \text{ minutes}$
3. cloudy
4. $30 \text{ days} \times 0.30 = 9 \text{ days}$
5. $30 \text{ days} \times 0.40 = 12 \text{ days}$
6. $\frac{400}{100} + \frac{75}{100} = \frac{475}{100} = 475\% = 4.75$

7. 0.25
8. 0.16
9. 0.06
10. 1.25
11. $3.61 \times 0.07 = 0.2527$
12. $0.9 - 0.14 = 0.76$
13. $2.69 + 8.7 = 11.39$
14. $\dfrac{44}{5} \div \dfrac{14}{3} = \dfrac{132}{15} \div \dfrac{70}{15} = \dfrac{132 \div 70}{1}$
    $= \dfrac{132}{70} = \dfrac{66}{35} = 1\dfrac{31}{35}$
15. $\dfrac{17}{5} \div \dfrac{47}{7} = \dfrac{119}{35} \div \dfrac{235}{35} = \dfrac{119}{235}$
16. $\dfrac{29}{2} \div \dfrac{17}{3} = \dfrac{87}{6} \div \dfrac{34}{6} = \dfrac{87}{34} = 2\dfrac{19}{34}$
17. $\$1.45 \times 1.15 = \$1.67$ (rounded)
18. 10 kg × 1,000 g/kg = 10,000 grams
19. 1.5 books/week × 6 weeks = 9 books
20. A liter is close to 1 quart; approximately 4 liters

Lesson Test 14

1. $(0.1) \times (0.5) = (0.05)$

   ```
         0.123
   ×     0.456
           11
          628
        ¹1 15
         5 0
         1 2
        ¹4 8
   0.0 5 6 0 8 8
   ```

2. $(3) \times (6) = (18)$

   ```
          3.0 5 2
   ×      6.1 9 3
         ¹9 ¹1 5 6
       ¹2²7 4 1
              5 8
         3 0 5 2
       18 3 1 2
   18.9 0 1 0 3 6
   ```

3. $(0.3) \times (0.3) = (0.09)$

   ```
        0.28 1
   ×    0.269
         729
        ²1 8
       ¹1 4 8 6
         2
       1 6 2
         4
   0.0 7 5 5 8 9
   ```

4. $(2) \times (5) = (10)$

   ```
          1.504
   ×      5.167
        ¹3 5 2 8
              7
        ¹3 0 2 4
              6
        ¹1 5 0 4
        2 5 2 0
              5
   7.7 7 1 1 6 8
   ```

5. 5
6. $\dfrac{1}{2}$
7. $\dfrac{38}{100} = \dfrac{19}{50}$
8. $\dfrac{2}{100} = \dfrac{1}{50}$
9. $\dfrac{900}{100} + \dfrac{40}{100} = \dfrac{940}{100} = 940\% = 9.40$
10. 16,000 mm
11. 350 hl
12. 10 dg
13. $\dfrac{5}{8} \div \dfrac{2}{3} = \dfrac{5}{8} \times \dfrac{3}{2} = \dfrac{15}{16}$
14. $\dfrac{1}{5} \div \dfrac{3}{8} = \dfrac{1}{5} \times \dfrac{8}{3} = \dfrac{8}{15}$
15. $\dfrac{1}{2} \div \dfrac{3}{5} = \dfrac{1}{2} \times \dfrac{5}{3} = \dfrac{5}{6}$
16. 10.4 gal × $1.639/gal = $17.05
17. 3.2 m × 39.37 in/m = 125.984 in
18. $6.00 × 1.20 = $7.20
19. $25.60 + $11.19 + $45.21 = $82.00
20. $82.00 × 1.11 = $91.02

Lesson Test 15

1. 0.07 m
2. 0.1 kl
3. 0.5 g
4. 0.25 hm
5. 80 ml
6. 9,500 cg
7. $(1) \times (0.007) = (0.007)$

$$\begin{array}{r} 1.465 \\ \times\ 0.007 \\ \hline {}^12\ 4\ 3 \\ 7\ 8\ 2\ 5 \\ \hline 0.010\ 2\ 5\ 5 \end{array}$$

8. $(200) \times (0.1) = (20)$

$$\begin{array}{r} 239.016 \\ \times\ 0.134 \\ \hline {}^{1}3\quad\ 2 \\ 82{}^{1}6{}^{1}044 \\ {}^{2}\ \ \ 1 \\ {}^{3}6\ 97\ 0\ 38 \\ {}^{1}2\ 3\ 9\ 0\ 1\ 6 \\ \hline 32.028\ 144 \end{array}$$

9. 0.06
10. 0.45
11. 1.30
12. 2.00
13. $34 \div 2 = 17$
14. $36 \div 9 = 4$
    $4 \times 2 = 8$
15. $100 \div 5 = 20$
    $20 \times 3 = 60$
16. $\dfrac{33}{8} \div \dfrac{6}{5} = \dfrac{\cancel{33}^{11}}{8} \times \dfrac{5}{\cancel{6}_2} = \dfrac{55}{16} = 3\dfrac{7}{16}$
17. $\dfrac{26}{3} \div \dfrac{1}{6} = \dfrac{26}{\cancel{3}} \times \dfrac{\cancel{6}^2}{1} = 52$
18. 1 kilogram
19. 500,000 mm
20. $35 \text{ tons} \times 1.08 = 37.8 \text{ tons}$

Lesson Test 16

1. $3.14(10 \text{ m})^2 = 314 \text{ m}^2$
2. $3.14(20 \text{ m}) = 62.8 \text{ m}$
3. $3.14(2.5 \text{ in})^2 = 19.625 \text{ in}^2$
4. $3.14(5 \text{ in}) = 15.7 \text{ in}$
5. 0.061 m
6. 0.004 kg
7. 130 ml
8. $0.07 \times 0.04 = 0.0028$
9. $72.3 \times 0.9 = 65.07$
10. $204 \times 0.11 = 22.44$
11. $\dfrac{17}{2} \div \dfrac{9}{4} = \dfrac{17}{\cancel{2}} \times \dfrac{\cancel{4}^2}{9} = \dfrac{34}{9} = 3\dfrac{7}{9}$
12. $\dfrac{19}{6} \div \dfrac{5}{3} = \dfrac{19}{\cancel{6}_2} \times \dfrac{\cancel{3}}{5} = \dfrac{19}{10} = 1\dfrac{9}{10}$
13. $\dfrac{9}{10} \div \dfrac{2}{5} = \dfrac{9}{\cancel{10}_2} \times \dfrac{\cancel{5}}{2} = \dfrac{9}{4} = 2\dfrac{1}{4}$
14. $9 \text{ m} + 4 \text{ m} + 9 \text{ m} + 4 \text{ m} = 26 \text{ m}$
15. $4.5 \text{ ft} + 4.5 \text{ ft} + 4.5 \text{ ft} + 4.5 \text{ ft} = 18 \text{ ft}$
16. $6.2 \text{ in} + 3 \text{ in} + 6.2 \text{ in} + 3 \text{ in} = 18.4 \text{ in}$
17. $3.14(6 \text{ ft})^2 = 113.04 \text{ ft}^2$
18. $3.14(5 \text{ ft}) = 15.7 \text{ ft}$
19. $8 \text{ yd} + 8 \text{ yd} + 8 \text{ yd} + 8 \text{ yd} = 32 \text{ yd}$
20. $2 \text{ hm} \times 100 \text{ m/hm} = 200 \text{ m}$

Unit Test II

1. $$\begin{array}{r} 8.61 \\ \times\ 0.7 \\ \hline 1\ 4 \\ 5\ 6\ 2\ 7 \\ \hline 6.027 \end{array}$$

2. $$\begin{array}{r} 1.4 \\ \times 2.6 \\ \hline 2 \\ 1\ 6\ 4 \\ 2\ 8 \\ \hline 3.64 \end{array}$$

3. $$\begin{array}{r} 0.18 \\ \times 0.95 \\ \hline 1\ 4 \\ 7\ 5\ 0 \\ 9\ 2 \\ \hline 0.17\ 1\ 0 \end{array}$$

4. $\begin{array}{r} 0.106 \\ \times\,0.352 \\ \hline 212 \\ {}^1530 \\ 318 \\ \hline 0.037312 \end{array}$

5. $\begin{array}{r} 2.605 \\ \times\,6.279 \\ \hline {}^5\\ {}^118445 \\ {}^4\\ {}^114235 \\ 1\\ {}^134210 \\ 12630 \\ \hline 16.356795 \end{array}$

6. $\begin{array}{r} 3.191 \\ \times\,4.260 \\ \hline 5\;\;0 \\ {}^1{}^28646 \\ 1\\ 6282 \\ 3\\ {}^112464 \\ \hline 13.59\,3660 \end{array}$

7. 0.01
8. 0.10
9. 1.00
10. $\dfrac{5}{100} = \dfrac{1}{20}$
11. $\dfrac{48}{100} = \dfrac{12}{25}$
12. $\dfrac{16}{100} = \dfrac{4}{25}$
13. $0.25 = 25\%$
14. $0.40 = 40\%$
15. $0.50 = 50\%$
16. $0.75 = 75\%$
17. $\dfrac{600}{100} + \dfrac{25}{100} = \dfrac{625}{100} = 625\% = 6.25$
18. 0.2 m
19. 8,100,000 cg

20.  electricity
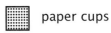 ingredients
paper cups
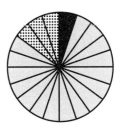

21. $\$35 \times 0.80 = \$28$
22. $3.14(3\text{ m})^2 = 28.26\text{ m}^2$
23. $3.14(6\text{ m}) = 18.84\text{ m}$
24. $\dfrac{27}{5} \times \dfrac{21}{5} = \dfrac{567}{25} = 22\dfrac{17}{25}$
25. $\dfrac{7}{3} \times \dfrac{5}{9} = \dfrac{35}{27} = 1\dfrac{8}{27}$
26. $\dfrac{15}{2} \div \dfrac{5}{4} = \dfrac{\cancel{15}^{\,3}}{\cancel{2}} \times \dfrac{\cancel{4}^{\,2}}{\cancel{5}} = 6$
27. $\dfrac{20}{3} \div \dfrac{2}{5} = \dfrac{{}^{10}\cancel{20}}{3} \times \dfrac{5}{\cancel{2}} = \dfrac{50}{3} = 16\dfrac{2}{3}$
28. $0.88 \times 50 = 44$ questions
29. $\$135.00 \times 1.16 = \$156.60$
30. $10.16 + 9.25 + 10.16 + 9.25 = 38.82$ in

Lesson Test 17

1. $\begin{array}{r} 0.002 \\ 2\overline{\smash{)}0.004} \\ \underline{4} \\ 0 \end{array}$

2. $\begin{array}{r} 0.002 \\ \times\,2 \\ \hline 0.004 \end{array}$

3. $\begin{array}{r} 0.22 \\ 8\overline{\smash{)}1.76} \\ \underline{160} \\ 16 \\ \underline{16} \\ 0 \end{array}$

4. $\begin{array}{r} 0.22 \\ \times\,8 \\ \hline 1 \\ 166 \\ \hline 1.76 \end{array}$

5. $\phantom{4\overline{)}}0.07$
    $4\overline{)0.28}$
    $\phantom{4\overline{)}}\underline{28}$
    $\phantom{4\overline{)0.2}}0$

6. 0.07
    $\underline{\times4}$
    0.28

7. $\phantom{5\overline{)}}0.2$
    $5\overline{)1.0}$
    $\phantom{5\overline{)}}\underline{10}$
    $\phantom{5\overline{)1.}}0$

8. 0.2
    $\underline{\times5}$
    1.0

9. $\phantom{7\overline{)}}0.5$
    $7\overline{)3.5}$
    $\phantom{7\overline{)}}\underline{35}$
    $\phantom{7\overline{)3.}}0$

10. 0.5
    $\underline{\times7}$
    3.5

11. $\phantom{3\overline{)}}0.003$
    $3\overline{)0.009}$
    $\phantom{3\overline{)0.00}}\underline{9}$
    $\phantom{3\overline{)0.00}}0$

12. 0.003
    $\underline{\times3}$
    0.009

13. $3.14(1.2 \text{ mi})^2 = 4.5216 \text{ mi}^2$
14. $3.14(2.4 \text{ mi}) = 7.536 \text{ mi}$
15. $40 \text{ in} + 30 \text{ in} + 40 \text{ in} + 30 \text{ in} = 140 \text{ in}$
16. $7.9 \text{ m} + 3.2 \text{ m} + 7.9 \text{ m} + 3.2 \text{ m} = 22.2 \text{ m}$
17. $6\frac{1}{4} + 4\frac{3}{4} + 6\frac{1}{4} + 4\frac{3}{4} = 20\frac{8}{4} = 22 \text{ in}$
18. $\$40.95 \div 7 = \$5.85$
19. $71.6 \text{ mi} \div 4 = 17.9 \text{ mi}$
20. $3.5 \text{ hr} + 6.9 \text{ hr} + 4.3 \text{ hr} = 14.7 \text{ hr}$
    $14.7 \text{ hr} \times \$7.10/\text{hr} = \$104.37$

Lesson Test 18

1. $\phantom{0.004\overline{)}}121{,}000$
    $0.004\overline{)484.000}$
    $\phantom{0.004\overline{)}}\underline{400{,}000}$
    $\phantom{0.004\overline{)0}}84{,}000$
    $\phantom{0.004\overline{)0}}\underline{80{,}000}$
    $\phantom{0.004\overline{)000}}4{,}000$
    $\phantom{0.004\overline{)000}}\underline{4{,}000}$
    $\phantom{0.004\overline{)00000}}0$

2. $\phantom{\times}121{,}000$
    $\underline{\times0.004}$
    $\phantom{\times}484.000$

3. $\phantom{0.18\overline{)}}201$
    $0.18\overline{)36.18}$
    $\phantom{0.18\overline{)}}\underline{3600}$
    $\phantom{0.18\overline{)000}}18$
    $\phantom{0.18\overline{)000}}\underline{18}$
    $\phantom{0.18\overline{)0000}}0$

4. 201
    $\underline{\times 0.18}$
    1608
    201
    $\overline{36.18}$

5. $\phantom{6\overline{)2}}3.7$
    $6\overline{)22.2}$
    $\phantom{6\overline{)}}\underline{18.0}$
    $\phantom{6\overline{)00}}42$
    $\phantom{6\overline{)00}}\underline{42}$
    $\phantom{6\overline{)000}}0$

6. 3.7
    $\underline{\times6}$
    14
    182
    $\overline{22.2}$

7. $\phantom{9\overline{)}}0.0006$
    $9\overline{)0.0054}$
    $\phantom{9\overline{)0.00}}\underline{54}$
    $\phantom{9\overline{)0.000}}0$

8. 0.0006
    $\underline{\times9}$
    0.0054

9. $3.14(3.5 \text{ in})^2 \approx 38.465 \text{ in}^2$
10. $3.14(7 \text{ in}) \approx 21.98 \text{ in}$
11. $13 \times 13 = 169$

12. $6 \times 6 = 36$
13. $10 \times 10 = 100$
14. $11 \text{ ft} + 11 \text{ ft} + 12 \text{ ft} = 34 \text{ ft}$
15. $9 \text{ in} + 12 \text{ in} + 15 \text{ in} = 36 \text{ in}$
16. $6.9 \text{ m} + 5.4 \text{ m} + 4.1 \text{ m} = 16.4 \text{ m}$
17. $\$19 \div 0.25 = 76$ quarters
18. $5.4 \text{ lb} \div 0.45 \text{ lb/batch} = 12$ batches
19. $13.9 + 13.9 + 13.9 + 13.9 = 55.6 \text{ m}$
20. $3{,}160 \text{ g} \times \dfrac{1 \text{ kg}}{1{,}000 \text{ g}} = 3.16 \text{ kg}$

Lesson Test 19

1. $0.19G = 38$
   $G = 38 \div 0.19$
   $G = 200$
2. $0.19(200) = 38$
   $38 = 38$
3. $0.008Y = 2$
   $Y = 2 \div 0.008$
   $Y = 250$
4. $0.008(250) = 2$
   $2 = 2$
5. $0.6G = 24$
   $G = 24 \div 0.6$
   $G = 40$
6. $0.6(40) = 24$
   $24 = 24$
7. $0.11Y = 55$
   $Y = 55 \div 0.11$
   $Y = 500$
8. $0.11(500) = 55$
   $55 = 55$
9.
```
         1,003
006.)6 018.
       6,000
          18
          18
           0
```
10.
```
    1,003
   ×0.006
    6.018
```

11.
```
      0.2
   15)3.0
      30
       0
```
12.
```
      1 5
    ×0.2
     3.0
```
13. $8 \text{ m} \times 4 \text{ m} = 32 \text{ m}^2$
14. $3.9 \text{ ft} \times 3.9 \text{ ft} = 15.21 \text{ ft}^2$
15. $5.1 \text{ in} \times 2.5 \text{ in} = 12.75 \text{ in}^2$
16. $0.45J = 90 \text{ hr}$
    $J = 90 \text{ hr} \div 0.45$
    $J = 200 \text{ hours}$
17. $200 \text{ hr} - 90 \text{ hr} = 110 \text{ hr}$
18. $0.75G = 12$
    $G = 12 \div 0.75$
    $G = 16$ gifts
19. $\$308.00 \div 8 \text{ people} = \$38.50/\text{person}$
20. $5{,}000 \text{ g} = 5 \text{ kg}$
    $50 \text{ kg} > 5 \text{ kg}$

Lesson Test 20

1.
```
         55
    5.)275.
       250
        25
        25
         0
```
2.
```
     5 5
   ×0.5
       2
    2 5
    27.5
```
3.
```
          21.25
   32.)680.00
        640.00
         40.00
         32.00
          8.00
          6.40
          1.60
          1.60
             0
```

LESSON TEST 20 - LESSON TEST 21

4.
$$\begin{array}{r} 21.25 \\ \times\ 0.32 \\ \hline \text{1 1 1} \\ 4240 \\ \text{1} \\ 6365 \\ \hline 6.8000 \end{array}$$

5.
$$\begin{array}{r} 680 \\ 04.\overline{)2720.} \\ \underline{2400} \\ 320 \\ \underline{320} \\ 0 \end{array}$$

6.
$$\begin{array}{r} 680 \\ \times 0.04 \\ \hline 3 \\ 2420 \\ \hline 27.20 \end{array}$$

7.
$$\begin{array}{r} 15 \\ 28.\overline{)420.} \\ \underline{280} \\ 140 \\ \underline{140} \\ 0 \end{array}$$

8.
$$\begin{array}{r} 0.28 \\ \times 15 \\ \hline 1 \\ 140 \\ 28 \\ \hline 4.20 \end{array}$$

9.
$$\begin{array}{r} 0.056 \\ 35\overline{)1.960} \\ \underline{1750} \\ 210 \\ \underline{210} \\ 0 \end{array}$$

10.
$$\begin{array}{r} 0.056 \\ \times\ \ 35 \\ \hline \text{1 3} \\ 250 \\ \text{1} \\ 158 \\ \hline 1.960 \end{array}$$

11.
$$\begin{array}{r} 4{,}270 \\ 002.\overline{)8{,}540.} \\ \underline{8{,}000} \\ 540 \\ \underline{400} \\ 140 \\ \underline{140} \\ 0 \end{array}$$

12.
$$\begin{array}{r} 4\ 2\ 7\ 0 \\ \times 0.002 \\ \hline 1 \\ 8\ 4\ 4\ 0 \\ \hline 8.5\ 4\ 0 \end{array}$$

13. $0.12Q = 0.36$
    $Q = 0.36 \div 0.12$
    $Q = 3$
    $0.12(3) = 0.36;\ 0.36 = 0.36$

14. $2.5W = 0.75$
    $W = 0.75 \div 2.5$
    $W = 0.3$
    $2.5(0.3) = 0.75;\ 0.75 = 0.75$

15. $\frac{17}{3} \div \frac{6}{7} = \frac{17}{3} \times \frac{7}{6} = \frac{119}{18} = 6\frac{11}{18}$

16. $23\frac{10}{10} - 12\frac{1}{10} = 11\frac{9}{10}$

17. $\frac{4}{\cancel{5}} \times \frac{\cancel{10}^2}{11} = \frac{8}{11}$

18. $10\ \text{in} \times 6\ \text{in} = 60\ \text{in}^2$

19. $27.3\ \text{mi} \div 9.1\ \text{mi/part} = 3\ \text{parts}$

20. $\$66.36 \div \$3.16/\text{item} = 21\ \text{items}$

Lesson Test 21

1.
$$\begin{array}{r} 21.325 = 21.33 \\ 4.\overline{)85.300} \\ \underline{80000} \\ 5300 \\ \underline{4000} \\ 1300 \\ \underline{1200} \\ 100 \\ \underline{80} \\ 20 \end{array}$$

LESSON TEST 21 - LESSON TEST 22

2. $0.245 = 0.25$
    $16\overline{)3.930}$
    $\underline{3.200}$
    730
    $\underline{640}$
    90
    $\underline{80}$
    10

3. $2.066 = 2.07$
    $09.\overline{)18.600}$
    $\underline{18000}$
    600
    $\underline{540}$
    60
    $\underline{54}$
    6

4. $7.4545 = 7.\overline{45}$
    $11.\overline{)82.0000}$
    $\underline{770000}$
    50000
    $\underline{44000}$
    6000
    $\underline{5500}$
    500
    $\underline{440}$
    60
    $\underline{55}$
    5

5. $44.0909 = 44.\overline{09}$
    $22.\overline{)970.0000}$
    $\underline{8800000}$
    900000
    $\underline{880000}$
    20000
    $\underline{19800}$
    200
    $\underline{198}$
    2

6. $1.166 = 1.1\overline{6}$
    $6\overline{)7.000}$
    $\underline{6000}$
    1000
    $\underline{600}$
    400
    $\underline{360}$
    40
    $\underline{36}$
    4

7. $30X = 0.8$
    $X = 0.8 \div 30$
    $X = 0.02\frac{2}{3}$

8. $0.5Y = 0.123$
    $Y = 0.123 \div 0.5$
    $Y = 0.24\frac{3}{5}$

9. $2.3F = 6.41$
    $F = 6.41 \div 2.3$
    $F = 2.78\frac{16}{23}$

10. $65 - 0.45 = 64.55$
11. $7.03 + 0.2 = 7.23$
12. $200 - 0.02 = 199.98$
13. $1/2(12 \text{ ft} \times 18 \text{ ft}) = 108 \text{ ft}^2$
14. $1/2(5 \text{ in} \times 12 \text{ in}) = 30 \text{ in}^2$
15. $1/2(10.2 \text{ m} \times 4.1 \text{ m}) = 20.91 \text{ m}^2$
16. $51.84 \text{ m}^2 \div 5.4 \text{ m} = 9.6 \text{ m}$
17. $\$25.45 \times 1.20 = \$30.54$
18. $3.14(8 \text{ in})^2 = 200.96 \text{ in}^2$
19. grams
20. $31 \text{ kg} \times 1,000 \text{ g/kg} = 31,000 \text{ g}$
    $31,000 \text{ g} - 310 \text{ g} = 30,690 \text{ g}$

Lesson Test 22

1. $0.09G + 0.38 = 0.425$
    $0.09G = 0.425 - 0.38$
    $0.09G = 0.045$
    $G = 0.045 \div 0.09 = 0.5$

2. $0.09(0.5) + 0.38 = 0.425$
    $0.045 + 0.38 = 0.425$
    $0.425 = 0.425$

3. $0.4X + 3.6 = 6.4$
    $0.4X = 6.4 - 3.6$
    $0.4X = 2.8$
    $X = 2.8 \div 0.4 = 7$

4. $0.4(7) + 3.6 = 6.4$
    $2.8 + 3.6 = 6.4$
    $6.4 = 6.4$

5. $1.5X + 4 = 4.165$
    $1.5X = 4.165 - 4$
    $1.5X = 0.165$
    $X = 0.165 \div 1.5 = 0.11$

6. $1.5(0.11) + 4 = 4.165$
    $0.165 + 4 = 4.165$
    $4.165 = 4.165$

7. $0.944 = 0.94$
    $9\overline{)8.500}$
    $\underline{8100}$
    400
    $\underline{360}$
    40
    $\underline{36}$
    4

8. $4.032 = 4.03$
    $4\overline{)16.130}$
    $\underline{16000}$
    130
    $\underline{120}$
    10
    $\underline{8}$
    2

9. $8.971 = 8.97$
    $07\overline{)62.800}$
    $\underline{56000}$
    6800
    $\underline{6300}$
    500
    $\underline{490}$
    10
    $\underline{7}$
    3

10. 1.8 m
11. 0.549 kl
12. 45 mg
13. $2.8 \text{ ft} \times 2.8 \text{ ft} \times 2.8 \text{ ft} = 21.952 \text{ ft}^3$

14. $\frac{5}{4} m \times \frac{5}{4} m \times \frac{5}{4} m = \frac{125}{64} m^3 = 1\frac{61}{64} m^3$
15. $3 \text{ in} \times 22 \text{ in} = 66 \text{ in}^2$
16. $22 \text{ in} + 4 \text{ in} + 22 \text{ in} + 4 \text{ in} = 52 \text{ in}$
17. $\frac{1}{2}(4 \text{ m} \times 6 \text{ m}) = 12 \text{ m}^2$
18. $5.4 \text{ m} + 4.2 \text{ m} + 6 \text{ m} = 15.6 \text{ m}$
19. $\$1.69 \times 6 = \$10.14$
    $\$10.14 \times 1.03 = \$10.44$
20. $0.45D + 1.35 = 23.85$
    $0.45D = 23.85 - 1.35$
    $0.45D = 22.50$
    $D = 22.50 \div 0.45 = \$50$

Lesson Test 23

1. $3 \div 10 = 0.30 = 30\%$
2. $1 \div 7 = 0.14\frac{2}{7} = 14\frac{2}{7}\%$
3. $2 \div 3 = 0.66\frac{2}{3} = 66\frac{2}{3}\%$
4. $5 \div 6 = 0.83\frac{1}{3} = 83\frac{1}{3}\%$
5. $1 \div 13 = 0.07\frac{9}{13} = 7\frac{9}{13}\%$
6. $8 \div 9 = 0.88\frac{8}{9} = 88\frac{8}{9}\%$
7. $0.9R + 0.5 = 0.77$
    $0.9R = 0.77 - 0.5$
    $0.9R = 0.27$
    $R = 0.27 \div 0.9 = 0.3$
8. $0.9(0.3) + 0.5 = 0.77$
    $0.27 + 0.5 = 0.77$
    $0.77 = 0.77$
9. $5.2833 = 5.28\overline{3}$
    $6\overline{)31.7000}$
    $\underline{300000}$
    17000
    $\underline{12000}$
    5000
    $\underline{4800}$
    200
    $\underline{180}$
    2

10. $303.33 = 303.3$
     $3.\overline{)910.00}$
     $\underline{90000}$
     1000
     $\underline{900}$
     100
     $\underline{90}$
     10
     $\underline{9}$
     1

11. $4.722 = 4.7\overline{2}$
     $09.\overline{)42.500}$
     $\underline{36000}$
     6500
     $\underline{6300}$
     200
     $\underline{180}$
     20
     $\underline{18}$
     2

12. $8 \text{ in} \times 9 \text{ in} \times 5 \text{ in} = 360 \text{ in}^3$

13. $3.4 \text{ m} \times 2.1 \text{ m} \times 1 \text{ m} = 7.14 \text{ m}^3$

14. $\frac{1}{2} \text{ ft} \times \frac{1}{2} \text{ ft} \times \frac{3}{4} \text{ ft} = \frac{3}{16} \text{ ft}^3$

15. $\frac{35}{50} = \frac{70}{100} = 70\%$

16. $1 \div 8 = 0.12\frac{1}{2} = 12\frac{1}{2}\%$

17. $\$5.65/\text{lb} \times 0.60 \text{ lb} = \$3.39$

18. $45 \times 0.80 = 36$ done
     $45 - 36 = 9$ toys left

19. $\$11.20 \times 1.20 = \$13.44$
     $\$20.00 - \$13.44 = \$6.56$

20. $100 \text{ ft}^2 \div 10 \text{ ft} = 10 \text{ ft}$

Unit Test III

1. 0.002
    $4\overline{)0.008}$
    $\underline{8}$
    0

2. 0.002
    $\underline{\times 4}$
    0.008

3. 0.05
    $6\overline{)0.30}$
    $\underline{30}$
    0

4. 0.05
    $\underline{\times 6}$
    0.30

5. 430
    $2.\overline{)860.}$
    $\underline{800}$
    60
    $\underline{60}$
    0

6. 430
    $\underline{\times 0.2}$
    86.0

7. 0.07
    $09.\overline{)00.63}$
    $\underline{63}$
    0

8. 0.07
    $\underline{\times 0.09}$
    0.0063

9. 16
    $21.\overline{)336.}$
    $\underline{210}$
    126
    $\underline{126}$
    0

10. 16
    $\underline{\times 2.1}$
    $1\,16$
    $\underline{2\,2}$
    33.6

11. $1,810$
    $005.\overline{)9,050.}$
    $\underline{5,000}$
    $4,050$
    $\underline{4,000}$
    50
    $\underline{50}$
    0

UNIT TEST III - LESSON TEST 24

12. $\begin{array}{r} 1,8\stackrel{4}{1}0 \\ \times 0.005 \\ \hline 5\,0\,5\,0 \\ \hline 9.0\,5\,0 \end{array}$

13. $6.3W = 11.34$
    $W = 11.34 \div 6.3$
    $W = 1.8$

14. $6.3(1.8) = 11.34$
    $11.34 = 11.34$

15. $0.7X + 0.08 = 17.58$
    $0.7X = 17.58 - 0.08$
    $0.7X = 17.5$
    $X = 17.5 \div 0.7 = 25$

16. $0.7(25) + 0.08 = 17.58$
    $17.5 + 0.08 = 17.58$
    $17.58 = 17.58$

17. $2 \text{ in} \times 10 \text{ in} = 20 \text{ in}^2$

18. $10 \text{ in} + 5.5 \text{ in} + 10 \text{ in} + 5.5 \text{ in} = 31 \text{ in}$

19. $1.4 \text{ m} \times 0.7 \text{ m} = 0.98 \text{ m}^2$

20. $1.4 \text{ m} + 0.7 \text{ m} + 1.4 \text{ m} + 0.7 \text{ m} = 4.2 \text{ m}$

21. $\frac{1}{2}(3 \text{ ft} \times 6 \text{ ft}) = 9 \text{ ft}^2$

22. $2 \text{ ft} + 4 \text{ ft} + 6 \text{ ft} = 12 \text{ ft}$

23. $\begin{array}{r} 12.233 \approx 12.23 \\ 6\overline{\smash{)}73.400} \\ \underline{60000} \\ 13400 \\ \underline{12000} \\ 1400 \\ \underline{1200} \\ 200 \\ \underline{180} \\ 20 \end{array}$

24. $\begin{array}{r} 15.166 \approx 15.1\overline{6} \\ 3\overline{\smash{)}45.500} \\ \underline{30000} \\ 15500 \\ \underline{15000} \\ 500 \\ \underline{300} \\ 200 \\ \underline{180} \\ 20 \end{array}$

25. $\begin{array}{r} 0.87\frac{4}{8} = 0.87\frac{1}{2} \\ 8\overline{\smash{)}7.00} \\ \underline{640} \\ 60 \\ \underline{56} \\ 4 \end{array}$

26. $9 \div 10 = 0.90 = 90\%$

27. $2 \div 7 = 0.28\frac{4}{7} = 28\frac{4}{7}\%$

28. $1 \div 3 = 0.33\frac{1}{3} = 33\frac{1}{3}\%$

29. $3 \div 4 = 0.75 = 75\%$

30. $8 \text{ ft} \times 9 \text{ ft} \times 7.2 \text{ ft} = 518.4 \text{ ft}^3$

Lesson Test 24

1. $\frac{85}{100} = \frac{17}{20}$

2. $\frac{16}{100} = \frac{4}{25}$

3. $\frac{2}{10} = \frac{1}{5}$

4. $\frac{5}{1000} = \frac{1}{200}$

5. $2 \div 9 = 0.22\frac{2}{9} = 22\frac{2}{9}\%$

6. $4 \div 5 = 0.80 = 80\%$

7. $1 \div 2 = 0.50 = 50\%$

8. $2 \div 3 = 0.66\frac{2}{3} = 66\frac{2}{3}\%$

9. $0.75X = 7.5$
   $X = 7.5 \div 0.75 = 10$

10. $0.75(10) = 7.5$
    $7.5 = 7.5$

11. $1.4X + 5 = 6.82$
    $1.4X = 6.82 - 5$
    $1.4X = 1.82$
    $X = 1.82 \div 1.4 = 1.3$

12. $1.4(1.3) + 5 = 6.82$
    $1.82 + 5 = 6.82$
    $6.82 = 6.82$

13. $\begin{array}{r} 13.33\frac{1}{3} \\ 3\overline{)40.00} \\ \underline{3000} \\ 1000 \\ \underline{900} \\ 100 \\ \underline{90} \\ 10 \\ \underline{9} \\ 1 \end{array}$

14. $\begin{array}{r} 2.43\frac{7}{11} \\ 11\overline{)26.80} \\ \underline{2200} \\ 480 \\ \underline{440} \\ 40 \\ \underline{33} \\ 7 \end{array}$

15. $1 + 5 + 9 = 15$
    $15 \div 3 = 5$

16. $2 + 2 + 5 + 7 = 16$
    $16 \div 4 = 4$

17. $75 \text{ lb} \times 1.30 = 97.5 \text{ lb}$

18. $45 \text{ in} + 31 \text{ in} + 45 \text{ in} + 31 \text{ in} = 152 \text{ in}$

19. $3.14(60 \text{ ft})^2 = 11{,}304 \text{ ft}^2$

20. $3.5 \text{ km} \times 1{,}000 \text{ m/km} = 3{,}500 \text{ m}$

Lesson Test 25

1. $8 + 13 + 17 + 17 + 19 + 19 + 19 = 112$
   mean $= 112 \div 7 = 16$
   median $= 17$
   mode $= 19$

2. $70 + 80 + 80 + 90 + 100 = 420$
   mean $= 420 \div 5 = 84$
   median $= 80$
   mode $= 80$

3. $8 + 8 + 9 + 10 + 15 = 50$
   mean $= 50 \div 5 = 10$
   median $= 9$
   mode $= 8$

4. $3 + 3 + 6 + 7 + 11 = 30$
   mean $= 30 \div 5 = 6$
   median $= 6$
   mode $= 3$

5. $\frac{53}{100}$

6. $\frac{44}{100} = \frac{11}{25}$

7. $23 \div 40 = 0.57\frac{1}{2} = 57\frac{1}{2}\%$

8. $4 \div 6 = 0.66\frac{2}{3} = 66\frac{2}{3}\%$

9. $9X + 3.4 = 7.9$
   $9X = 7.9 - 3.4$
   $9X = 4.5$
   $X = 4.5 \div 9 = 0.5$

10. $9(.5) + 3.4 = 7.9$
    $4.5 + 3.4 = 7.9$
    $7.9 = 7.9$

11. $10 \times 10 = 100$

12. $3 \times 3 \times 3 = 27$

13. $1 \times 1 \times 1 \times 1 \times 1 \times 1 = 1$

14. $0.9 \text{ m} + 1 \text{ m} + 0.9 \text{ m} + 1 \text{ m} = 3.8 \text{ m}$

15. $0.9 \text{ m} \times 0.8 \text{ m} = 0.72 \text{ m}^2$

16. $9.7 \text{ ft} + 9.7 \text{ ft} + 9.7 \text{ ft} + 9.7 \text{ ft} = 38.8 \text{ ft}$

17. $9.7 \text{ ft} \times 9.7 \text{ ft} = 94.09 \text{ ft}^2$

18. $2 + 2 + 3 + 3 + 4 + 4 + 4 + 6 + 8 = 36$
    average, or mean $= 36 \div 9 = 4$

19. most often, or mode $= 4$

20. middle, or median $= 4$

Lesson Test 26

1. $\frac{10}{1000} = \frac{1}{100}$

2. $\frac{300}{1000} = \frac{3}{10}$

3. $\frac{700}{1000} = \frac{7}{10}$

4. $\frac{4}{10} = \frac{2}{5}$

5. $\dfrac{4}{10} = \dfrac{2}{5}$

6. $\dfrac{7}{10}$

7. $3+7+8+8+9 = 35$
   mean $= 35 \div 5 = 7$
   median $= 8$
   mode $= 8$

8. $9+9+12+13+17 = 60$
   mean $= 60 \div 5 = 12$
   median $= 12$
   mode $= 9$

9. $\dfrac{5}{10} = \dfrac{1}{2}$

10. $\dfrac{92}{100} = \dfrac{23}{25}$

11. $\dfrac{9}{100}$

12. $\dfrac{825}{1000} = \dfrac{33}{40}$

13. $1 \div 3 = 0.33\dfrac{1}{3} = 33\dfrac{1}{3}\%$

14. $7 \div 8 = 0.87\dfrac{1}{2} = 87\dfrac{1}{2}\%$

15. $3.14(6.4 \text{ ft}) = 20 \text{ ft}$

16. $200 \times 0.16 = 32$ people

17. 500 meters

18. $(8+4-6+3) \times 2 = 18$

Lesson Test 27

1. endpoint
2. length; width
3. ray
4. line segment
5. length; width
6. c
7. d
8. e
9. b
10. a
11. $\overrightarrow{XY}$
12. $\overleftrightarrow{QR}$ or $\overleftrightarrow{RQ}$
13. $1+1+1+2+5 = 10$
    mean $= 10 \div 5 = 2$
14. median $= 1$

15. mode $= 1$
16. $2.05 \times 1.3 = 2.665$
17. $1.43 \div 0.11 = 13$
18. $1.75 + 2.25 = 4$
19. $\dfrac{1}{2}$
20. $\dfrac{50}{100} = \dfrac{1}{2}$
    $8 \div 2 = 4$
    $4 \times 1 = 4$ chores done

Lesson Test 28

1. point
2. plane
3. plane
4. equal
5. congruent
6. similar
7. ↔
8. ~
9. ∞
10. →
11. •
12. —
13. ▱
14. ≅
15. 50%
16. 25%
17. 75%
18. $33\dfrac{1}{3}\%$
19. 20%
20. $66\dfrac{2}{3}\%$

Lesson Test 29

1. rays
2. $\dfrac{1}{4}$
3. 90
4. 360
5. vertex
6. similar
7. congruent

LESSON TEST 29 - UNIT TEST IV

8. length; width
9. ∠RGX or ∠XGR
10. $\frac{5}{10} = \frac{1}{2}$
11. $\frac{8}{100} = \frac{2}{25}$
12. $\frac{565}{1000} = \frac{113}{200}$
13. $\frac{\cancel{4}^2}{5} \times \frac{3}{\cancel{10}_5} = \frac{6}{25}$
14. $\frac{3}{24} + \frac{14}{24} = \frac{17}{24}$
15. $\frac{5}{8} - \frac{4}{8} = \frac{1}{8}$
16. $\frac{5}{150} = \frac{1}{30}$
17. 9 is most often, or mode
18. $16.34 + 37.2 + 45.5 + 61 = 160.04$ lb
    160.04 lb $\div 4 = 40.01$ lb

Lesson Test 30

1. 90°
2. acute
3. straight
4. obtuse
5. right
6. straight
7. obtuse
8. acute
9. 480%
10. 91%
11. 9%
12. $0.25 \times 300 = 75$
13. $1.50 \times 80 = 120$
14. $0.30 \times 75 = 22.5$
15. $\frac{2}{3} \div \frac{1}{8} = \frac{2}{3} \times \frac{8}{1} = \frac{16}{3} = 5\frac{1}{3}$
16. $\frac{4}{9} \div \frac{1}{3} = \frac{4}{\cancel{9}_3} \times \frac{\cancel{3}}{1} = \frac{4}{3} = 1\frac{1}{3}$
17. $15 \times 15 = 225$
18. $3.14(12.2 \text{ ft}) = 38.308$ ft

Unit Test IV

1. $\frac{52}{100} = \frac{13}{25}$
2. $\frac{8}{100} = \frac{2}{25}$
3. $\frac{3}{10}$
4. $\frac{575}{1000} = \frac{23}{40}$
5. $10 + 10 + 11 + 12 + 17 = 60$
   mean $= 60 \div 5 = 12$
   median $= 11$
   mode $= 10$
6. $8 + 13 + 17 + 17 + 19 + 19 + 19 = 112$
   mean $= 112 \div 7 = 16$
   median $= 17$
   mode $= 19$
7. $\frac{40}{280} = \frac{1}{7}$
8. $\frac{210}{280} = \frac{3}{4}$
9. $\frac{40 + 30}{280} = \frac{70}{280} = \frac{1}{4}$
10. $\overline{FG}$ or $\overline{GF}$
11. $\overleftrightarrow{RS}$ or $\overleftrightarrow{SR}$
12. $\overrightarrow{AW}$
13. ∠HJK or ∠KJH
14. ray
15. line segment
16. length; width
17. point
18. plane
19. congruent
20. equal
21. similar
22. vertex
23. 360
24. 90°
25. acute
26. straight
27. obtuse
28. c
29. e
30. f
31. b
32. d

UNIT TEST IV - FINAL TEST

33. g
34. a

Final Test

1. $1 \times 1 \times 1 \times 1 \times 1 \times 1 = 1$
2. $8 \times 8 = 64$
3. $10 \times 10 \times 10 = 1000$
4. 5,271.349
5.
$$\begin{array}{r} {}^{6}\!\!\not{7}.{}^{14}\!\!\not{5}\,{}^{1}\!2 \\ -\;1.\;8\;5 \\ \hline 5.\;6\;7 \end{array}$$
6.
$$\begin{array}{r} 6.0 \\ +5.28 \\ \hline 11.28 \end{array}$$
7.
$$\begin{array}{r} 3\,{}^{1}\!2.{}^{9}\!\!\not{0}\,{}^{1}\!\!\not{3}\not{4}\,{}^{1}\!1 \\ -0.\;5\;\;\;9\;6 \\ \hline 31.\;4\;\;\;4\;5 \end{array}$$
8.
$$\begin{array}{r} 2.49 \\ \times\,0.6 \\ \hline 2\,5 \\ 12\,4\,4 \\ \hline 1.4\,9\,4 \end{array}$$
9.
$$\begin{array}{r} 1.7 \\ \times\;\;3 \\ \hline 2 \\ 3\,1 \\ \hline 5.1 \end{array}$$
10.
$$\begin{array}{r} 0.004 \\ \times\,0.05 \\ \hline 0.00020 \end{array}$$
11. 1,300,000 cm
12. 0.25 g
13. 0.05
14. 0.65
15. $\dfrac{25}{100} = \dfrac{1}{4}$
16. $\dfrac{32}{100} = \dfrac{8}{25}$
17. $0.8 = 80\%$
18. $5 \div 6 = 0.83\dfrac{1}{3} = 83\dfrac{1}{3}\%$
19. $\dfrac{400}{100} + \dfrac{60}{100} = \dfrac{460}{100} = 4.60 = 460\%$
20. $\dfrac{78}{100} = \dfrac{39}{50}$
21. $\dfrac{3}{100}$
22.
$$\begin{array}{r} 3.325 \approx 3.33 \\ 4\,\overline{|\,13.300} \\ \underline{12\,00} \\ 130 \\ \underline{120} \\ 10 \\ \underline{8} \\ 2 \end{array}$$
23.
$$\begin{array}{r} 0.654 \approx 0.65 \\ 7\,\overline{|\,4.580} \\ \underline{4200} \\ 380 \\ \underline{350} \\ 30 \\ \underline{28} \\ 2 \end{array}$$
24.
$$\begin{array}{r} 65.66 \approx 65.\overline{6} \\ 0.6\,\overline{|\,39.400} \\ \underline{36000} \\ 3400 \\ \underline{3000} \\ 400 \\ \underline{360} \\ 40 \\ \underline{36} \\ 4 \end{array}$$
25.
$$\begin{array}{r} 0.733 \approx 0.\overline{73} \\ 0.03\,\overline{|\,.02200} \\ \underline{2100} \\ 100 \\ \underline{90} \\ 10 \\ \underline{9} \\ 1 \end{array}$$
26.
$$\begin{array}{r} 0.81\dfrac{9}{11} \\ 11\,\overline{|\,9.00} \\ \underline{880} \\ 20 \\ \underline{11} \\ 9 \end{array}$$

27. $\begin{array}{r}0.55\\9\overline{)5.00}\\\underline{450}\\50\\\underline{45}\\5\end{array}\;\frac{5}{9}$

28. $3.2X + 0.07 = 4.55$
    $3.2X = 4.55 - 0.07$
    $3.2X = 4.48$
    $X = 4.48 \div 3.2$
    $X = 1.4$

29. $3.2(1.4) + 0.07 = 4.55$
    $4.48 + 0.07 = 4.55$
    $4.55 = 4.55$

30. plane
31. line segment
32. point
33. length; width
34. ray
35. obtuse
36. acute
37. similar
38. 360
39. 90°
40. straight
41. congruent
42. $A = 3.14(3 \text{ ft})^2 = 28.26 \text{ ft}^2$
    $C = 3.14(6 \text{ ft}) = 18.84 \text{ ft}$
43. $\$3.50 + \$5 + \$5 + \$7 + \$8 = \$28.50$
    mean = $\$28.50 \div 5 = \$5.70$
    median = $\$5$
    mode = $\$5$
44. $\$45.60 \times 1.14 \approx \$51.98$
45. $\dfrac{1}{758}$

Word Problems 6

1. $3.45 + $1.99 + $6.59 + $12.98 = $25.01
   $25.01 − $2.50 = $22.51
   $50.00 − $22.51 = $27.49 left

2. Use a drawing to show their travels. The distances don't have to be to scale.

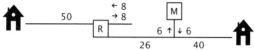

   50 mi + 26 mi + 40 mi = 116 miles from house to house
   50 mi + 8 mi + 8 mi = 66 miles past restaurant and back
   66 mi + 26 mi = 92 miles to turnoff
   92 mi + 6 mi + 6 mi = 104 miles back to main route
   104 mi + 40 mi = 144 miles total driven

3. 3.5" − 0.6" = 2.9"
   2.9" + 8.3" = 11.2"
   11.2" − 4.2" = 7" remaining

Word Problems 12

All money is rounded to hundredths at each step.

1. 8 × $8.40 = $67.20
   7 × $5.99 = $41.93
   $67.20 + $41.93 = $109.13 discountable
   $109.13 × 0.10 ≈ $10.91 (rounded)
   $109.13 − $10.91 = $98.22
   for discounted yarn
   2 × $2.50 = $5.00 for undiscounted
   $98.22 + $5.00 = $103.22 cost of yarn
   $103.22 × 1.13 = $116.64
   with tax and shipping (5% + 8% = 13%)

2. $5.99 × 5 = $29.95
   0.10 × $29.95 = 2.995 rounds to $3.00
   $29.95 − $3.00 = $26.95 discounted price
   $26.95 × 1.13 = $30.45
   with tax and shipping
   You could also figure the total cost of one skein and multiply by 5. Rounding may give a slightly different answer.

3. $2.50 × 1.13 ≈ 2.83
   for one skein (rounded)
   $2.83 × 1.5 ≈ $4.25 (rounded)
   You could have found the basic cost of 1.5 skeins first and then figured tax and shipping. Your answer may be slightly different because of rounding.

Word Problems 18

1. $\frac{1}{4} + \frac{1}{2} + \frac{3}{4} + \frac{1}{4} =$
   $\frac{1}{4} + \frac{2}{4} + \frac{3}{4} + \frac{1}{4} =$
   $\frac{7}{4}$ or $1\frac{3}{4}$ pizza left over
   $\frac{7}{4} \div \frac{1}{4} = \frac{7}{4} \times \frac{4}{1} = 7$ boys

2. 5.3" ÷ 10 = 0.53"
   water from average snow
   4.1" ÷ 5 = 0.82" water from wet snow

   If you are unsure whether to multiply or divide, check your answer to see if it makes sense. The actual snowfall in each case was less than what was needed to make an inch of water, so division yields a sensible answer.
   0.53" + 0.82" + 1.5" = 2.85"
   rounds to 2.9" of water

3. (2)3.14 × 3' = 18.84'
   circumference of small garden
   2 × 3' = 6' diameter of small garden
   3.14 × 12' = 37.68' circumference of garden with doubled diameter
   37.68' − 18.84' = 18.84'
   additional edging needed

   In real life this would probably be rounded to 19 or 20 feet.

4. $3.14 \times (3 \text{ ft})^2 = 28.26 \text{ ft}^2$ area
    of small garden
    28 ft^2 rounded
    $3.14 \times (6 \text{ ft})^2 = 113.04 \text{ ft}^2$ area
    of large garden
    113 ft^2 rounded
    $113 \text{ ft}^2 \div 28 \text{ ft}^2/\text{pack.} = 4.04$ or
    4 seed packets (rounded)
    When you have learned how to divide a decimal by a decimal, try this again using the unrounded areas.

Word Problems 24

1. $8 \text{ in} \times 12 \text{ in} = 96 \text{ in}^2$ area of paper
    $3.14(3 \text{ in})^2 = 28.26^2 \text{ in}$ area of one circle
    $3.14(2.5 \text{ in})^2 = 19.625 \text{ in}^2$ or 19.63 in^2
    rounded area of other circle
    $28.26 \text{ in}^2 + 19.63 \text{ in}^2 = 47.89 \text{ in}^2$
    used for circles
    $96 \text{ in}^2 - 47.89 \text{ in}^2 = 48.11 \text{ in}^2$ left over
    Look at the drawing to see why it is not possible to cut another circle with a 3" radius, even though there seems to be enough area.
    One circle has a diameter of 6", which leaves 2" distance to the edge of the paper.
    The other circle has a diameter of 5", which leaves a 3" distance to the edge of the paper.
    Neither space is enough for another circle with a 3" radius (6" diameter).

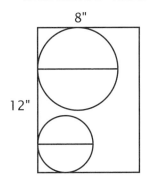

2. $\$6.50 \times 0.10 = \$.65$ is 10%
    of his hourly income
    $\$6.50 - 0.65 = \$5.85$
    hourly amount available to spend
    $\$25.35 + \$50.69 + \$85.96 =$
    $\$162$ total needed
    $\$162 \div \$5.85 = 27.69...$
    rounds to 28 hours

3. $0.3 \times T = 300$ blue toys
    $T = 300 \div 0.3$
    $T = 1000$ toys in all
    $\frac{1}{4} \times 1000 = 250$ green toys
    $300 + 250 = 550$ blue or green toys
    $1000 - 550 = 450$ remaining toys
    $\frac{1}{2} \times 450 = 225$ red toys
    $\frac{1}{2} \times 450 = 225$ yellow toys
    You could use decimals or fractions for this problem.

Symbols and Tables

U.S. CUSTOMARY MEASUREMENT

3 teaspoons (tsp) = 1 tablespoon (Tbsp)
2 pints (pt) = 1 quart (qt)
8 pints = 1 gallon (gal)
4 quarts = 1 gallon
12 inches (in) = 1 foot (ft)
3 feet = 1 yard (yd)
5,280 feet = 1 mile (mi)
16 ounces (oz) = 1 pound (lb)
2,000 pounds = 1 ton

METRIC MEASUREMENT

1,000 millimeters (mm) = 1 meter (m)
100 centimeters (cm) = 1 meter
10 decimeters (dm) = 1 meter
10 meters = 1 dekameter (dam)
100 meters = 1 hectometer (hm)
1,000 meters = 1 kilometer (km)

Replace meter with liter or gram for liquid and mass equivalents.

METRIC – U.S. EQUIVALENTS

1 meter ($\approx$ 39.37 in) $\approx$ 1.09 yd (36 in)
1 centimeter $\approx$ 0.4 inch
1 liter $\approx$ 1.06 quarts
1 kilogram $\approx$ 2.2 pounds
1 kilometer $\approx$ 0.6 mile
1 inch $\approx$ 2.5 cm or 1 inch = 2.54 cm
1 ounce $\approx$ 28 grams

SYMBOLS

=	equals
~	similar
$\cong$	congruent
$\approx$	approximately equal to
<	less than
>	greater than
%	percent
π	pi (approximately 22/7 or 3.14)
r^2	r squared, or r · r
'	foot
"	inch
∞	infinity
·	point
$\leftrightarrow$	line
$\rightarrow$	ray
$\angle$	angle
▱	plane
⦜	right angle
| |	absolute value
$0.\overline{3}$	repeating decimal

PERIMETER

rectangle, square, triangle, parallelogram – add the lengths of all the sides

CIRCUMFERENCE

circle $C = 2\pi r$

AREA
rectangle, square, parallelogram
$$A = bh \text{ (base times height)}$$
triangle
$$A = \frac{bh}{2} \text{ or } \frac{1}{2}bh$$
circle
$$A = \pi r^2$$

VOLUME
rectangular solid, cube
$$V = Bh \text{ (area of base times height)}$$

Glossary

A

absolute value - the value of a number without its sign, or the difference between a number and zero expressed as a positive number

acute angle - an angle with a measure greater than 0° and less than 90°

algebraic expression - a mathematical phrase that can contain numbers, variables, and operation symbols

angle - a geometric figure formed by two rays joined at their origins

area - the measure of the space covered by a plane shape, expressed in square units

average - a measure of center in a set of numbers; an average could be measured using a mean, median, or mode, but usually it refers to the mean

B–D

base - 1. a particular side or face of a geometric figure used to calculate area or volume; 2. a number that is raised to a power. For example, in the expression 34, the base is 3.

base ten - a number system based on ten, also called *decimal system*

base unit - a metric unit of measurement that can be modified by adding a prefix to represent fractions or multiples

centi - in the metric system, the Latin prefix representing one hundredth of the base unit

circumference - the distance around the outside of a circle

coefficient - a quantity placed before and multiplying the variable in an algebraic expression

Commutative Property - a property which states that the order in which numbers are added or multiplied does not affect the result

congruent - having exactly the same size and shape

conversion factor - a ratio equal to one that is used to convert measures; also called *unit multiplier*

cube - a solid with six congruent faces that meet at right angles

deci - in the metric system, a Latin prefix representing one tenth of the base unit

decimal (fraction) - a fraction written using a decimal point and place value

decimal point - a dot used to separate whole numbers and fractions; also used to separate dollars and cents

decimal system - a number system based on ten, also called base ten

deka - or *deca*- in the metric system, the Greek prefix representing ten of the base unit

denominator - the bottom number in a fraction, which shows the number of parts in the whole

diameter - a straight line passing through the center of a circle and touching both sides

Distributive Property of Multiplication over Addition - a property for multiplying a sum by a given factor

dividend - the number being divided

divisor - a number that is being divided into another

factor - (n) a whole number that multiplies with another to form a product; (v) to find the factors of a given product

fraction - a number indicating part of a whole

geometry - a branch of mathematics that deals with figures in space

gram - the basic unit of mass in the metric system

greatest common factor (GCF) - the greatest number that will divide evenly into two or more numbers

E–G

endpoint - 1. one of the starting or ending points of a line segment; 2. another name for the origin of a ray

equation - a mathematical statement that uses an equal sign to show that two expressions have the same value

equivalent - having the same value

estimate - a close approximation of an actual value

even number - any number that can be evenly divided by two

expanded notation - a way of writing numbers by showing each digit multiplied by its place value

exponent - a raised number that indicates the number of times a factor is multiplied by itself; also called power

exponential notation - a form of expanded notation where each place value is indicated by 10 with an exponent

H–L

hecto - in the metric system, the Greek prefix representing 100 of the base unit

height - the perpendicular distance from the base to the top of a figure

improper fraction - a fraction with a numerator greater than its denominator

inverse - opposite or reverse

kilo - in the metric system, the Greek prefix representing 1,000 of the base unit

least common multiple (LCM) - the least number that is a multiple of two or more other numbers

line - in geometry, a set of connected points that extends infinitely in two directions

line segment - a section of a line bounded by two endpoints

liter - the basic unit of liquid volume in the metric system

M–O

mass - the measure of the amount of matter in an object

mean - also called *average*; a measure of center found by dividing the sum of a set of values by the number of values

median - the middle value in a list of numbers when they are arranged in order from least to greatest

meter - the basic unit of linear measure in the metric system

metric system - a system of measurement based on ten

milli - in the metric system, the Latin prefix representing one thousandth of the base unit

mixed number - a number written as a whole number and a fraction

mode - in a data set, the item that appears the most often

negative number - a number less than zero

number line - a line on which every point corresponds to a real number

numerator - the top number in a fraction, which shows the number of parts being considered

obtuse angle - a triangle in which one of the angles is greater than 90 degrees

origin - 1. another name for the endpoint of a ray; 2. on a coordinate grid, the point at the intersection of the axes, generally identified by the ordered pair (0, 0)

P–R

parallel lines - lines in the same plane that do not intersect

percent - a ratio with 100 as the second part; shown with the symbol %

perimeter - the distance around a polygon

perpendicular lines - lines that form right angles when they intersect

pi - the Greek letter π, which represents an irrational number with an approximate value of 22/7 or 3.14

pie graph - also called *pie chart* or *circle graph*; a circular chart divided into sections that show parts of a whole amount

place value - the position of a digit which indicates its assigned value

place-value notation - a way of writing numbers that shows the place value of each digit

plane - a flat, two-dimensional surface that extends infinitely in all directions

point - a defined position in space that has no dimensions; represented with a dot

power - another name for an exponent; indicates the number of times a factor is multiplied by itself

probability - the likelihood of getting a desired outcome, given all possible outcomes

proper fraction - a fraction with a numerator less than its denominator

quadrilateral - a polygon with four sides

quadrant - one of the four sections of a Cartesian coordinate grid formed by the axes

quotient - the result when numbers are divided

radius - the distance from the center of a circle to its edge; in a regular polygon, the distance from the center to any vertex; in a sphere, the distance from the center to any point on the surface; plural is *radii*

rate - a ratio that compares quantities with different units of measure

ratio - the relationship between two values; can be written in fractional form

rational numbers - numbers that can be written as ratios or fractions, including decimals

ray - a geometric figure that starts at a definite point (called the origin) and extends infinitely in one direction

reciprocal - the number that, when multiplied by a given number, has a product of 1; multiplicative inverse

rectangle - a quadrilateral with two pairs of opposite parallel sides and four right angles

rectangular solid - a three-dimensional shape with six rectangular faces

repeating decimal - a decimal which eventually shows a repeating pattern of digits; the repeating digits can be shown under a line, called a *vinculum*

right angle - an angle measuring 90 degrees

rounding - replacing a number with another that has approximately the same value but is easier to use

"Rule of Four" - A Math-U-See method for finding the common denominator of two fractions

S–U

similar - having the same shape but a different size

simplify - to rewrite an expression as simply as possible; a simplified fraction will have a numerator and denominator with a single common factor of one

square - a quadrilateral in which the four sides are perpendicular and congruent

straight angle - an angle with a measure of 180 degrees

surface area - the sum of the areas of all the faces of a solid

symmetry - having congruent parts facing each other across an axis, with one the reverse of the other

term - a part of an algebraic expression which may be a number, a variable, or a product

unit - the place in a place-value system representing numbers less than the base

unit multiplier - a ratio equal to one that is used to convert measures; also called conversion factor

unknown - a specific quantity that has not yet been determined, usually represented by a letter

V–Z

variable - a value that is not fixed or determined, often representing a range of possible values

vertex - the endpoint shared by the two rays of an angle

volume - the number of cubic units that can be contained in a solid

whole numbers - the set of numbers that begin with zero and continue to count up by one (0, 1, 2, 3, …)

X-axis - the horizontal number line used as a reference on a Cartesian coordinate system

Y-axis - the vertical number line used as a reference on a Cartesian coordinate system

Master Index for General Math

This index lists the levels at which main topics are presented in the instruction manuals for *Primer* through *Zeta*. For more detail, see the description of each level at mathusee.com. (Many of these topics are also reviewed in subsequent student books.)

Addition
    facts .... Primer, Alpha
    multiple-digit .... Beta
Additive inverse .... Epsilon
Angles .... Zeta
Area
    circle .... Epsilon, Zeta
    parallelogram .... Delta
    rectangle .... Primer, Gamma, Delta
    square .... Primer, Gamma, Delta
    trapezoid .... Delta
    triangle .... Delta
Average .... Delta
Circle
    area .... Epsilon, Zeta
    circumference .... Epsilon, Zeta
    recognition .... Primer, Alpha
Circumference .... Epsilon, Zeta
Common factors .... Epsilon
Comparison of amount .... Primer, Beta
Composite numbers .... Gamma
Congruent .... Zeta
Counting .... Primer, Alpha
Decimals
    add and subtract .... Zeta
    change to percent .... Epsilon, Zeta
    divide .... Zeta
    from a fraction .... Epsilon, Zeta
    multiply .... Zeta
Divisibility rules .... Epsilon
Division
    facts .... Delta
    multiple-digit .... Delta
Estimation .... Beta, Gamma, Delta
Expanded notation .... Delta, Zeta
Exponential notation .... Zeta
Exponents .... Zeta
Factors .... Gamma, Delta, Epsilon
Fractions
    add and subtract .... Epsilon
    compare .... Epsilon

    divide .... Epsilon
    equivalent .... Gamma, Epsilon
    fractional remainders .... Delta, Epsilon
    mixed numbers .... Epsilon
    multiply .... Epsilon
    of a number .... Delta, Epsilon
    of one .... Delta, Epsilon
    rational numbers .... Zeta
    to decimals .... Epsilon, Zeta
    to percents .... Epsilon, Zeta
Geometry
    angles .... Zeta
    area ...... Primer, Gamma, Delta, Epsilon,
    .... Zeta
    circumference .... Epsilon, Zeta
    perimeter .... Beta
    plane .... Zeta
    points, lines, rays .... Zeta
    shape recognition .... Primer, Alpha
Graphs
    bar and line .... Beta
    pie .... Zeta
Inequalities .... Beta
Linear measure .... Beta, Epsilon
Mean, median, mode .... Zeta
Measurement equivalents .... Beta, Gamma
Metric measurement .... Zeta
Mixed numbers .... Epsilon
Money .... Beta, Gamma
Multiplication
    facts .... Gamma
    multiple-digit .... Gamma
Multiplicative inverse .... Epsilon
Number recognition .... Primer
Number writing .... Primer
Odd and even numbers .... Alpha, Gamma
Ordinal numbers .... Beta
Parallel lines .... Delta
Parallelogram .... Delta
Percent
    decimal to percent .... Epsilon, Zeta

ZETA   MASTER INDEX FOR GENERAL MATH   **227**

    fraction to percent............ Epsilon, Zeta
    greater than 100..................... Zeta
    of a number............................. Zeta
Perimeter.. Beta
Perpendicular lines............................. Delta
Pi .... Epsilon, Zeta
Pie graph... Zeta
Place value...
    Primer, Alpha, Beta, Gamma, Delta, Zeta
Prime factorization......................... Epsilon
Prime numbers .... Gamma
Probability .... Zeta
Rational numbers.............................. Zeta
Reciprocal....................................... Epsilon
Rectangle
    area.................. Primer, Gamma, Delta
    perimeter .... Beta
    recognition .... Primer, Alpha
Rectangular solid .... Delta
Regrouping............................. Beta, Gamma
Roman numerals................................ Delta
Rounding.................... Beta, Gamma, Delta
Sequencing .... Beta
Skip counting
    2, 5, 10 .... Primer, Alpha, Beta
    all... Gamma
Solve for unknown ....
    ..Primer, Alpha, Gamma, Delta, Epsilon,
    .... Zeta
Square
    area.................. Primer, Gamma, Delta
    perimeter .... Beta
    recognition .... Primer, Alpha
Statistics .... Delta, Zeta
Subtraction
    facts .... Alpha
    introduction .... Primer, Alpha
    multiple-digit................................ Beta
Tally marks .... Primer, Beta
Telling time ....Primer, Alpha, Beta
Thermometers and gauges.................. Beta
Trapezoid .... Delta
Triangle
    area.. Delta
    perimeter .... Beta
    recognition .... Primer, Alpha
Volume of rectangular solid .... Delta

Zeta Index

Topic	Lesson
Acute angle	30
Algebra-decimal inserts	4, 9
Algebraic expressions	19
Angles	29, 30
Area	
circle	16
rectangle and square	Student 19D
parallelogram	Student 20D
triangle	Student 21D
Average	Student 24D, 25
Base	Student 19D, 20D, 21D
Base ten	3
Circle	16
Circumference	16
Coefficient	19, 22
Common denominator	Student 4D
Congruent	28
Cube	1, Student 22D
Decimal point	3, 4, 10, 14, 17, 18, 20
Decimals	
add	4
change to percent	11, 23
defined	3
divide	17, 18, 20, 21
expanded notation	3
inserts	4, 9
multiply	9, 10, 14
rational numbers	24
subtract	5
Diameter	16
Dividend	17
Divisor	17
Endpoint	27
Equal	28
Equations, solving	19, 22
Expanded notation	2, 3
Exponential notation	2, 3
Exponents	1, 2
Factor	9
Fractions	
add	Student 4D, 8D
common denominator	Student 4D
divide	Student 12D, 13D, 14D, 15D
equivalent	Student 2D
fractional remainders	21
improper	Student 6D, 7D, 13D
mixed numbers	Student 6D, 7D, 8D, 9D, 11D, 13D, 15D
multiply	Student 10D, 11D
of a number	Student 1D
rational numbers	24
Rule of Four	Student 4D, 12D
simplify	Student 3D
subtract	Student 5D, 9D
to decimals	23
to percents	11, 23
Geometry	27, 28, 29, 30
Height	Student 19D, 20D, 21D
Infinite	27
Interstate Highway System	Student 27D, E
Inverse	19
Invert and multiply	Student 14D
Line	27
Line segment	27
Mean	25
Median	25
Mental math	Student 26D
Metric measurement	
conversions	8, 15
gram	6
liter	6
meter	6
prefixes and origin	6, 7
Mixed numbers	Student 6D, 7D, 8D, 9D, 11D, 13D, 15D
Mode	25
Money	3, 18, 20
Obtuse angle	30
Origin	27
Percent	
decimal to percent	11, 23
fraction to percent	11, 23
greater than 100	12
of a number	11
pie graphs	13
Perimeter	
parallelogram	Student 17D
rectangle, square	Student 16D
triangle	Student 18D
Pi	16
Pie graph	13

Topic .... Lesson

Place value 2
Plane 28
Point 27
Power of a number 1
Probability 26
Product 9
Quotient 17
Radius 16
Ratio 24
Rational numbers 24
Ray 27
Reciprocal .... Student 14D
Rectangle .... Student 16D, 19D
Rectangular solid (prism) .... Student 23D
Remainders 21
Rounding .... Student 11C – #15
Right angle 29
Similar 28
Solving for unknown 19, 22
Square .... 1, Student 16D, 19D
Statistics 25
Straight angle 30
Taxes 11, 12
Tips 11, 12
U.S. customary measurements 7
Vertex 29
Variable 22
Volume
 cube .... Student 22D
 rectangular solid (prism) .... Student 23D
Word problems
 multi-step 6, 12, 18, 24
 tips 1